The Planet Neptune

An Exposition and History

JOHN PRINGLE NICHOL

CAMBRIDGE
UNIVERSITY PRESS

CAMBRIDGE UNIVERSITY PRESS

Cambridge, New York, Melbourne, Madrid, Cape Town,
Singapore, São Paolo, Delhi, Tokyo, Mexico City

Published in the United States of America by Cambridge University Press, New York

www.cambridge.org
Information on this title: www.cambridge.org/9781108038331

© in this compilation Cambridge University Press 2011

This edition first published 1848
This digitally printed version 2011

ISBN 978-1-108-03833-1 Paperback

CAMBRIDGE LIBRARY COLLECTION

Books of enduring scholarly value

Physical Sciences

From ancient times, humans have tried to understand the workings of the world around them. The roots of modern physical science go back to the very earliest mechanical devices such as levers and rollers, the mixing of paints and dyes, and the importance of the heavenly bodies in early religious observance and navigation. The physical sciences as we know them today began to emerge as independent academic subjects during the early modern period, in the work of Newton and other 'natural philosophers', and numerous sub-disciplines developed during the centuries that followed. This part of the Cambridge Library Collection is devoted to landmark publications in this area which will be of interest to historians of science concerned with individual scientists, particular discoveries, and advances in scientific method, or with the establishment and development of scientific institutions around the world.

The Planet Neptune

J. P. Nichol (1804–59), astronomer and political economist, was Regius Professor of Astronomy at the University of Glasgow. He brought astronomy to a non-scientific audience through his enthusiastic public lectures and astronomy books. His works include the popular *Views of the Architecture of the Heavens* (1837; also reissued in this series) in which he supported the nebular hypothesis, which in modified form is the model of star formation most widely accepted today. Neptune was (in 1846) the first planet to be discovered by mathematical prediction rather than empirical observation, and in this book, first published in 1855, Nichol describes that discovery to a lay readership. Part 1 is an exposition of the then current view of the solar system and the research and discoveries which led to that view; Part 2 is dedicated to Neptune; while the third part explains the controversies over the planet's discovery.

Cambridge University Press has long been a pioneer in the reissuing of out-of-print titles from its own backlist, producing digital reprints of books that are still sought after by scholars and students but could not be reprinted economically using traditional technology. The Cambridge Library Collection extends this activity to a wider range of books which are still of importance to researchers and professionals, either for the source material they contain, or as landmarks in the history of their academic discipline.

Drawing from the world-renowned collections in the Cambridge University Library, and guided by the advice of experts in each subject area, Cambridge University Press is using state-of-the-art scanning machines in its own Printing House to capture the content of each book selected for inclusion. The files are processed to give a consistently clear, crisp image, and the books finished to the high quality standard for which the Press is recognised around the world. The latest print-on-demand technology ensures that the books will remain available indefinitely, and that orders for single or multiple copies can quickly be supplied.

The Cambridge Library Collection will bring back to life books of enduring scholarly value (including out-of-copyright works originally issued by other publishers) across a wide range of disciplines in the humanities and social sciences and in science and technology.

PLATE XI.

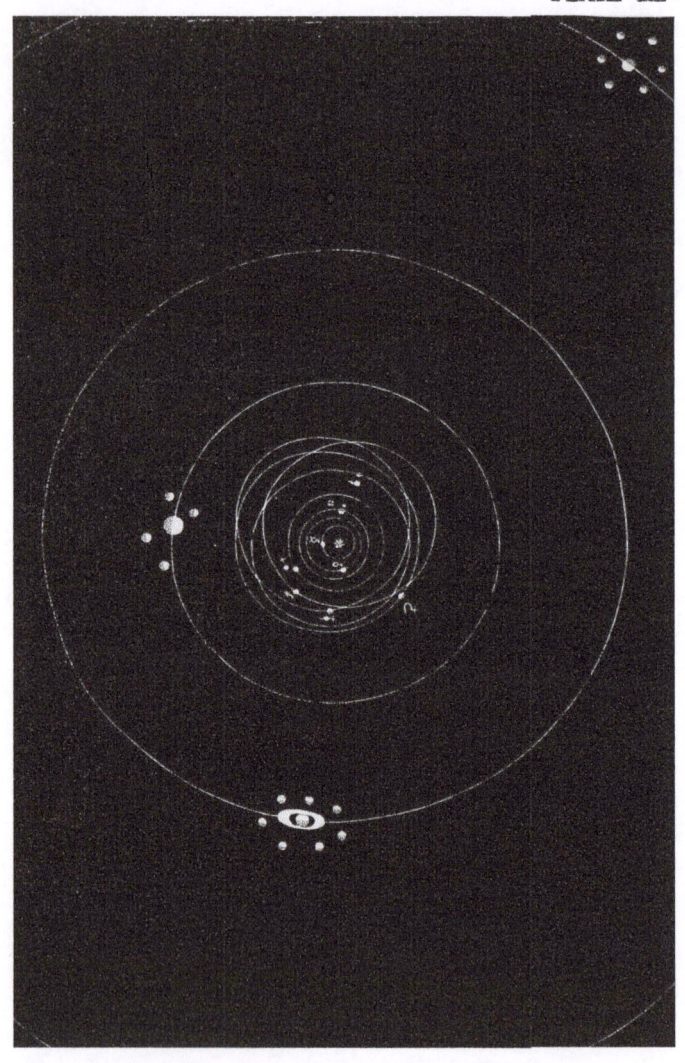

ARRANGEMENT OF THE SOLAR SYSTEM

THE

PLANET NEPTUNE:

AN

EXPOSITION AND HISTORY.

BY

J. P. NICHOL, LL.D.,

PROFESSOR OF ASTRONOMY IN THE UNIVERSITY OF GLASGOW.

AUTHOR OF

THE ARCHITECTURE OF THE HEAVENS," "CONTEMPLATIONS ON THE SOLAR SYSTEM,"
" THOUGHTS ON IMPORTANT POINTS RELATING TO THE SYSTEM OF THE WORLD,"
AND " THE STELLAR UNIVERSE," ETC. ETC.

JOHN JOHNSTONE,
15, PRINCES STREET, EDINBURGH; AND
26, PATERNOSTER ROW, LONDON.

MDCCCXLVIII.

Printed by JOHN JOHNSTONE, 104, High Street, Edinburgh.

TO

THE DIRECTORS

OF

THE EDINBURGH PHILOSOPHICAL INSTITUTION..

GENTLEMEN,—

In inscribing to you the following treatise, I only perform a distinct duty; for it contains the substance of a Lecture delivered before the Members of your excellent Body, and is published at your request. I cannot, however, let slip the opportunity of expressing my deep interest in an Institution with which I can now reckon the connection of many years, and my anxious wishes for its prolonged success.

I have the honour to be,

Yours very faithfully,

J. P. NICHOL.

OBSERVATORY, GLASGOW,
Nov. 1, 1847.

PREFACE.

I MAY be permitted to say as a Preface to the following Essay, that I hope my scientific readers will recollect the nature of its aim. It is no very difficult matter to write popular treatises on science, on the principle which one sees too frequently adopted, and which is wholly inconsistent with any attempt at accuracy or philosophic truth; but it is not easy to give an exposition of a matter so recondite as that which has just engaged me, so that without the falsification and degradation of science, it may yet be generally appreciated. The most successful of such attempts in this country is undoubtedly that of Mr. AIRY, in his remarkable work entitled "GRAVITATION;" but even that singular clearness and firmness of apprehension which distinguish this eminent mathematician, have not always preserved him from something approaching to obscurity. There are portions of the present Essay—especially that important part immediately succeeding page 91—

in which I scarcely expect that my effort will be found wholly satisfactory; and, indeed, there are difficulties still belonging to that portion of the subject, which will not be cleared up to our entire satisfaction previous to the full analytical discussion and accurate estimate of the actual relations between Neptune and Uranus.

The latter part of the Essay, having much to do with personal disputes, is one that may be supposed to have given the author of it pain. I confess, however, that I have experienced little hesitation or delicacy in treating those questions as I have done. The entire history is, on the whole, so creditable, and the occasions on which one would be inclined to find fault have reference to men in every way so able to bear a little criticism, that really there is no temptation to the individual who desires to narrate the plain unvarnished truth, at any time to colour or depart from it. I have had no object, either in the Exposition or the History, other than the wish to enable my countrymen, generally, to understand all the incidents connected with one of the most remarkable discoveries hitherto recorded in the Annals of Science.

J. P. N.

CONTENTS.

LIST OF PLATES.

NEPTUNE.

I HAVE delayed complying with the request made
to me during last winter by the Directors of the
Philosophical Institution of Edinburgh, that I
might have it in my power to lay before the public,
not merely an account of the leading features of
the history of the discovery of the new Planet,
but something more approaching to a statement
of the place which the relationships of that orb
must fix for it, amid the complex mechanism of our
System; and also of those subsidiary inquiries
which owe to it their origin. It cannot indeed
be alleged, even yet, that the positive theory of
Neptune is much more than begun; but the sagacity of the great Men who, in their ever-memorable adventure into that region of pure thought,

descried the necessity of the Planet as the complement of our System, and the industry of some excellent Observers, have already accomplished enough, to enable us to speak with no inconsiderable precision, alike of its individual and peculiar habitudes, and of the modifications which its existence imposes on our former ideas of the magnitudes and connections of its companion Globes.

I believe that, in respect of the larger proportion of those who will peruse this Essay, I might safely have assumed, the general disposition of the Solar System, as well as the grand Law of Gravitation, to be familiar and understood characteristics of Nature—an expression in a form and language no longer foreign to the intelligent classes of society, of the manner of that Order which pervades the mechanism of the starry Universe: but probably the employment of a few moments in recalling succinctly the character of that Law, especially in the two separate phases in which it appeared to its Immortal Discoverer, will not, at this very outset of our exposition, be altogether thrown away; inasmuch as the distinctest idea of it must be borne along with us, if we would tread with pleasure and understanding even through

one step of the course we have adventured on.
I shall attempt, therefore, although at the risk of
discoursing on what requires little elucidation,
to present, in the first part of my Essay, a suc-
cinct, but yet sufficiently detailed account of the
plan of our Planetary Scheme, and of the Relations
which constitute it—notwithstanding its variety
and complexity—one compact and perfect System.
Should the reader succeed in thoroughly realizing
this preliminary exposition, the subsequent history of the discovery of Neptune will be perused
with ease; and we shall be in a condition to adjust intelligently, and without bias by unworthy
partialities, the claims and relations of the illustrious Inquirers who have accomplished this great
feat in Science.

PART FIRST.

PICTURE OF THE SOLAR SYSTEM.

I.

THE System of the Sun and Planets, as we first descry it, is alike simple and majestic. Resting in one portion of Space, from which his lustre is diffused through the profundities which environ him, our magnificent Luminary ranks in glory, and corresponds in destiny, with the myriads of the fixed Stars. Around this orb, which illumines, cherishes, and upholds them, those smaller worlds—of which our Earth is one— roll, with admirable and unwavering regularity, at divers distances and in stated periods. Nearest of all, as shown in the *Frontispiece*, is placed the planet Mercury: next in the order of remoteness, we descry the brilliant Venus: then our

Earth, with its Moon: then Mars: then that complex group of small planets,—remarkable, through their extreme diminutiveness, and also because the orbits in which they move, so nearly approach each other, that one mean distance might almost be taken as indicating the position of them all: behind these lies Jupiter, with his satellites—the noblest and most beautiful among the secondary constituents of our system: then Saturn with his Moons and remarkable Ring: and finally the planet Uranus, which, until the advent of these later revelations, we deemed to be at the outer limit of planetary existence—the remotest of the regular globes attendant on our Sun. If we compare it with the Earth, how stupendous the magnitude of this System! but as part of the fathomless Universe, it is only as a speck or islet amid Space! Although that planet, Uranus, is removed from the central Luminary by a space nineteen times larger than that which divides us from the Sun, the entire span of its orbit so dwindles away beside the immense gulfs that intervene between Star and Star, that, from the nearest of these orbs which light up the skies of midnight, it would all seem covered over by an ordinary spider's thread! But how exquisite the arrangement of this system! How still and imperturbable the unwinding of its diverse motions! Even as a speck of crystal—scarcely seen in the crevice of the amorphous mass in which it lies—unfolds, nevertheless, the fullest perfections of that Art through which all inorganic matter approaches to form and order,—the thoughtful eye, as it pursues the relations of these minute worlds, will discern the presence of the Wisdom that built up the august Heavens—the energy of the grand Laws that uphold them all.

II.

THE plan and disposition of our Planetary System, as I have now described them, were revealed to mankind by the immortal labours of COPERNICUS. But in the epoch subsequent to that of the illustrious Pole, farther inquiries were started, and profounder and much more arduous problems placed before Astronomers. The question speedily became, not merely, In what order are these various orbs located, and in what manner do they move?—but, Can that mechanism as a whole, be reconciled with any known mechanical principles—are these arrangements of the Heavens indicative of the presence of Laws whose efficacy we discern among the changes more immediately around us? The nature of this inquiry will be most readily understood, through a simple narrative of the solution which, from the genius of SIR ISAAC NEWTON, it so speedily received. Prior to the time of this great Englishman—chiefly, in fact, by the sagacity of GALILEO and JOHN KEPLER—one grand principle, characteristic of the habitudes of all bodies in motion, had been satisfactorily evolved and apprehended. It was this : *Every body, on receiving an impulse or other disposition to move, will move onwards in a straight line, in the direction of the impelling force, provided it continue undisturbed by any other force.* Now, on a first glance at the system of the planetary motions, it is manifest, that these orbs do not move in straight lines, but in *continuous curves;* and therefore that their existing condition cannot be explained, by the mere supposition of their having been once set in motion by a special impulse or force: nay, it follows immediately, that, to this primal moving power, the action of some other disturbing force must be added, so that the phenomena be explained: and a little deeper consideration will show, that this second force cannot be one that has merely acted at a certain epoch and then ceased; but that, on the contrary, its energy is unceasing or *continuous.* —Let us reflect for a moment on the condition of a body moving, as on next page, in a *circle.*

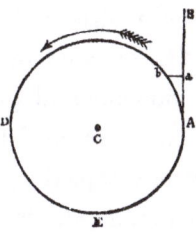

It may easily be made plain that the body A, revolving through the circle A, D, E, around C, cannot
be conceived, at any point of its course, to move,
merely in obedience to impulses previously given
it, and free from other restraint. When at A,
for instance—whatever these former impulses—its
tendency, were it free, would be to dart along
the line A B, exactly like a stone released from
a sling; so that its natural course would be A *a*,
instead of the circular one A *b*; and we cannot
imagine it pursuing freely the latter path, unless
some other power act on it *continually*—drawing it
from the right line towards which it inclines,
downwards at every moment, in the direction *a b*.
There is thus needed a steady and unintermitting
deflecting power—capable of bending straight lines
like A B, into such curves as constitute the orbits
of the Planets—to render in anywise intelligible
the order of our Solar System: and our great

Countryman at once discerned that, the inquiry
into the nature of this second and necessary Force
constitutes the fundamental problem of Astronomy.—The leading characteristics of this second
Power, did not long escape the persistent sagacity
of NEWTON. He discerned, almost on his earliest
glance, that it must be a power directed towards the
centre of the Sun: so that the condition of each
Planet seemed to be this:—It is sustained in its
actual course, by the conflict and equilibrium of
two forces—one, the result of the first impulse, or
cause of its motion, which would induce it to fly
off into space; and another, a restraining power
in uninterrupted action, which *of itself* would
cause the body to pass towards the centre of the
Sun. Animated by this centre-seeking tendency
on the one hand, and the *centrifugal*, or centre-
flying tendency on the other, the planet Jupiter
rolls his unceasing round in a path subject to no
accident, and, in as far as this goes, incapable of
change: and as it is the same central authority,
—merely changing in strength, along with the distance of the body from the Sun—which affects all
the orbs of our System—from Mercury ever hid
within the solar rays, to the remote and inhos-
pitable Uranus; its mechanism appears as simple

as its plan—the result of only two of those primary ordinances, that guide the phenomena of Motion.—Many profound inquiries—deeper indeed than Man has the power to resolve—are inevitably suggested by the scheme whose simplicity has thus burst on us. What, for instance, is this *power* which binds all these orbs to the Sun's centre? ATTRACTION—GRAVITATION—these are the names we have bestowed upon it: but what mean they? what the idea we are entitled to attach to them? Our positive knowledge, indeed, goes no farther than this—that the *order* of the planetary motions can be expressed by a certain Law; that the places each Orb occupies successively, in the course of its revolution, admit of being so joined together; or denoted in their totality, by the central influence we have assumed. Is Gravitation, then, only a name for this succession? Or rather, is it not the apparition of some primal Force or POWER, which by the order it bestows on material events—manifests its character, and influence? *This*, at least, we know—far as our ken at present extends, whether we follow those singular and majestic activities among the double, triple, and quadruple stars ; or adventure still farther, even among those awful adyta where the mightier groups of the Stellar Universe take on their forms and significance,—the influence of this august Law ever meets our gaze, preserving through all that exists, its matchless order, and in silence elaborating those more perfect developments, which will occupy the Durations to come.

III.

THE account of the mechanism of our System, which I have just presented, is very far from being the correct or completed one. We shall find it necessary, indeed, for the sake of distinctness, or at least highly advantageous, to begin our investigation by realizing even this imperfect view; but it conveys little conception of the actual structure of a Scheme, which is as complex as that representation was simple; and which, so far from presenting the orbit of any planet as the balance of only *two* tendencies, derives it from the equilibrium of sympathies alike multiplied and changeable—sympathies, uniting by some invisible bond not the planets with the Sun merely, but every atom of matter within our System with every other; and diffusing throughout its entire domains a unity so perfect and comprehensive, that it most resembles the wondrous vitality of some organized and sensitive framework. The principle of that unity is as follows: The Attraction or Gravitation which, by drawing the planets towards the Sun, restrains them within a definite curvilineal orbit, is not, as we have just supposed, confined to the Sun, or exemplified by his action, alone. Speaking of it in the usual, and, I believe, also in the most correct manner, *it is an energy or force exemplified by every particle of matter.* So universal is this energy, and so unlimited its sphere, that even the slightest pebble on the sea-shore sends forth an effective notice of its being, to the remotest orb amongst the profound recesses of space: nay, raise that pebble from where it lies, alter its relative position by a quantity however small, and your act is felt throughout the unfathomed Universe! Before the vision of a harmony so unbroken, of a mechanism so vast, so complex, and yet so perfect, the thoughtful mind cannot avoid being penetrated with awe! Yes! this Fabric is indeed the offspring of Omniscience, and under the care of a Government which protects alike the majestic and the feeble—where the very hairs of our heads are numbered!—There are two special points in the Solar System, as we must now view it, to which I would here solicit particular attention.

I. The simplicity of the conditions under which each planet performs its motions has now vanished utterly away. Not only is it attracted by the Sun; but—since Gravitation is universal—by every other orb having part among its arrangements. Refer once more to the Chart in the Frontispiece, and observe the case—say of the planet Venus. That orb indeed still obeys the influences I formerly adverted to, viz., her own tendency to fly off into space, and that counteracting tendency in virtue of which she would, if unchecked, move towards the centre of the Sun: but besides; she is pulled by all the other Planets, in most various directions; and not only with various, but ever-varying forces, inasmuch as each is dragging her towards itself with an energy depending on its size and distance; and that distance is—through those motions which are the life of the system—always changing. It is true indeed, and will be readily manifest; that of all these powers whose office it is to give the planet a curvilineal instead of its natural rectilineal path, the energy of the SUN is by far the most important; for, as I have said, one element in the efficacy of the power exercised by any body over another, is the *magnitude* of the deflecting body: and it will be seen in Plate II., which illustrates the relative massiveness, &c., of the bodies composing our system, that the Sun immensely transcends the aggregate of all the Planets, which indeed are rather his subordinate attendants.* The Sun's action, then, *still determines in the main* the path of Venus and these other orbs; and the action of the neighbouring planets only *disturbs* this path, or impresses on it certain changeable irregularities. For instance, if the Sun alone had acted, Venus would revolve around the central luminary in a perfectly true, or accurate geometrical curve, viz., the Ellipse: and that Ellipse may still, as below, be deemed the *normal* or *mean* path of that Planet,—her real one vibrating around it, as in the fanciful curve in the same diagram—a curve which is not reproduced

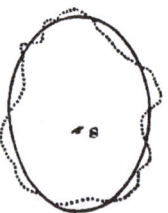

* In this plate the line A B at the top is meant to represent the diameter of the Sun, while the relations of the other Orbs are shown directly. Neptune is omitted—not yet being fully determined. It may, however, be taken as approximating to Uranus.

PLATE II.

URANUS.

SATURN.

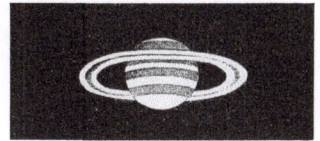

JUPITER.

- VESTA.
- JUNO.
- PALLAS.
- CERES.
- MARS.
- EARTH.
- VENUS.
- MERCURY.

A

B

year after year, but which will pass through as
many variations as the course of ages can witness,
in the relative positions of the constituents of our
multiplex scheme. To determine that normal
ellipse in the case of any Planet, might have been
no arduous task; but how different with these
deviations from it—these ever-shifting *perturbations!* The fidelity with which he can trace and
estimate directions and quantities so mutable and
evanescent, is the highest attainment of the Astronomer;—an object indeed most worthy of those
toils that have occupied him since the lifetime of
NEWTON. Many a midnight has been consumed
in the arduous pursuit; and success in the most
difficult portions of the enterprise, has conferred
on several names a distinction that will never die.
Before me in greatest brilliance, arises that illustrious triumvirate—EULER, CLAIRAUT, D'ALEM-
BERT; on one side of them the great family of
the BERNOUILLIS: on the other, our own MAC-
LAURIN, and in later times those two chiefs,
seated on twin summits—LAGRANGE and LAPLACE;
from whom, again, the mantle has descended
and fallen on men most worthy to wear it,—a
company goodly in number, and ardent with

hope; among whom, however, the eye easily discerns — as ever conspicuous — those illustrious fellow-victors in the race, ADAMS and LEVERRIER.

II. The modification of chief moment, however, which this completed idea of Gravitation impresses on our conception of the Solar System, is this : No part of it can be regarded as *separate* or independent,—each orb being *united* by ties which cannot be severed, *with all the others*, in regard of the chief features of its being. So long as we regard the Sun as the sole seat of this attractive energy, any Planet might be removed from the system without other consequence than the disappearance of a Star from our skies:—annihilate Jupiter, for instance, and all others would, in that case, roll onward precisely as before, in courses perfectly determinate and geometrically regular.—But if the merest pebble on a beach cannot be removed, without remote stars discerning its change of position, how entirely impossible this unobserved annihilation of Jupiter! Nay, so far from any planet being independent, or, as I should rather say, severed from relations with the great Uni-

verse (for such, in the present case, is the meaning
of *independence*),—if we restrict our thoughts to its
motions or orbit, and to its absolute magnitude,
that Planet may be said to be *determined by, or to be
the result of, the concurrence of the habitudes of all
the others.* In the case of Jupiter, for instance,
suppose him concealed from my sense of vision, by
being bereft of his power to reflect the radiance
of the Sun;—so long, nevertheless, as he rolls in
that orbit, his presence must be revealed to intellectual vision, by his influence on the motions of
the orbs within his reach: yea, the irregular and
then unaccounted-for motions of Saturn, would
point towards the reality of that Planet; and
could we aright interpret them, they would unfold alike the magnitude of the force that has
caused them—or, what is the same thing, the
mass or weight of Jupiter—and also the variability
or changeableness of that force, which, being interpreted, signifies the orbit and motions of Jupiter—
even as *unerringly and unceasingly as the movements
of the mysterious needle tell of the direction and intensity of the Influence toward which it turns.* Let the
regular line on page 28, for instance, be Saturn's
orbit, in so far as that is due to the Sun, and the
unbroken irregular line, his real orbit. Now, if

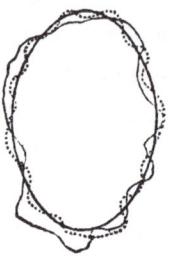

the dotted line indicates the orbit that Saturn
ought to assume in consequence of the action of
all known planets except Jupiter; then the discrepancy between these two last, viz.:—the irregular unbroken and dotted lines; would point to
the foreign influence; and nothing save the requisite
skill could be awanting to enable us to reach,
through these unexplained deviations, the reality
and definite habitudes of Jupiter. *Definite,* I say,
for in consequence of our thorough knowledge, not
only of the reality of Gravitation, but of the precise mode of its dependence on the magnitude and
distance of the bodies exercising its power—there
is now no room within the range of such an Inquiry, for *chance,* or toleration for *vagueness:* to
reach the fixed and definite conclusion in regard

of any point to which inquiry is directed, needs *skill alone.*—This hypothesis of a successful search for an opaque and invisible Jupiter, is not, however, merely an illustrative one;—it is, on the contrary, the equivalent of many investigations, that are far from strange in Astronomy. For instance, we read in works on that science—often indeed to the marvel of those who know not our technical proceedings—of the WEIGHTS of the several Planets ; which are spoken of with as little hesitation as if they had been weighed in our terrestrial balances. Now, these are deduced in this wise: A planet being disturbed by its neighbours, its movements are watched and scrutinized, while the deflecting bodies occupy with respect to it *all variety of positions.* The effect due to each, is thus separated from the total effect; and from the amount of its influence, the mass or weight of each planet is inferred : nay, speaking more generally, this process of determining important physical attributes of these orbs, by inference from their attractive influences, has constituted the principal branch of physical Astronomy from Newton's time until our own. Shall I dwell, in illustration, on the lights thus obtained regarding the *shape* of our own Earth ? That body, like all globes in rotation, is a spheroid flattened at the poles, or bulging out—as if into an attached *ring*—at the equator. The determination of the precise shape of the Earth, being of consummate consequence to geographical science, several of the greatest of our European States have instituted vast triangulations—some of them in remote regions—that, by the sure methods of the Surveyor, this character of our globe might be defined. Results of high accuracy have undoubtedly been reached by such labours ; but with all their importance, they are not superior to intelligence, which on the same subject has reached us from the *Moon.* The motions of that orb are not what they would be, had the Earth been *a perfect sphere.* Certain deviations are caused by the influence of our equatorial protuberance: and these deviations —measured by our modern instruments, whose precision approaches to the marvellous—enable us, by inverse reasoning, to determine with undoubted exactness, how far the Earth deviates from a regular globe. It is certainly unnecessary to accumulate illustrations of a point concerning which the whole modern history of Astronomy is most rich and emphatic : let me conclude, therefore, with an allusion to that recent grand generalization of the deeply-lamented BESSEL, which has unfolded the probability of opaque orbs large as these spangling suns, revolving with a Sirius or a Procyon,* around some common centre ; thus appalling our Imagination, already bewildered amid the excess of visible glories, by the unexpected and overwhelm-

* The method of this discovery, and the reasoning on which it reposes, are too remarkable to be permitted to escape with this transient notice. Every star in the heavens, it is known, has an apparent proper motion;—sometimes that motion being altogether *apparent*, and in other cases *real*. While studying, with his usual care and sagacity, the changes of place exhibited through a long series of years, by SIRIUS and PROCYON, Bessel saw that they evinced a real proper motion, of a very singular description. In both cases the Star OSCILLATES—that is, it moves backwards and forwards like a pendulum. Like Newton of old, Bessel marked this as an *unnatural* motion—one opposed to every known principle in mechanics. His first inference, then, was, that the oscillating motion is, in reality, a motion in a *circle*, or other curvilinear path, turned nearly *edgeways* towards us, and therefore taking on the appearance of an *oscillation*. But the investigation could not stop here. If the path is curvilinear,—that must be owing to a *continuous deflecting force*, as Newton held in regard of the Planets: and since no Body is visible around which either Sirius or Procyon can be supposed to revolve: the conclusion is, that the seat of this great attractive influence is *opaque*—concealed for ever from the sense of vision. The inference is so logical that we may not repudiate it: Surely, however, it is most wonderful—bringing the strange intelligence, that those brilliant orbs are, after all, but one class or species, amid the luxuriance of Creation!

ing idea, that the objects which light reveals, may, after all, constitute but one class, or one special though splendid scheme, amid this profusion of created magnificence !

IV.

I MUST now ask my readers to enter with some
minuteness on a consideration of the different
classes into which we separate these perturbations; that so they may thoroughly understand the
technical nature of our knowledge of the system.
I shall expose what I wish to be rightly apprehended, under three distinct heads : be it at once
remarked, however, that although we *class* these
perturbations, there is no difference, or ground for
classification among their *causes*. The principle
of this separation simply is, that for the sake of
convenience, we count up, in the first place, the
more obvious and considerable disturbances;
then those which are somewhat more evanescent, and extend probably over a larger period;
and lastly, those which enclose within their
wide range the whole conditions of our system—even in its relations with the fixed stars,

—stretching our view of the harmony of the various orbs, onwards through innumerable centuries.

I. The first and most palpable description of variations, or perturbations, is what we term PERIODICAL. These depend on the directions in which the different bodies lie, in regard of each other; and they simply affect each body's place in its orbit. An irregularity of this kind, *does not affect the ellipse or the species of curve in which any body moves;* but it causes that body to be either *before or behind its natural place, in that curve.* The method of taking account of such, in our calculations of the place of a Planet at some future time, is extremely simple;—we determine, by another description of inequalities, in what ellipse or curve the body must then be moving, and the laws of the periodical inequality readily determine in what place of the ellipse the body ought to appear. These perturbations, too, have generally very short periods; and although they may often be of much less absolute importance than deflections which stretch over a very wide range,—they are, nevertheless, all-important, when we are required to study the course

of the planet only *through a short period of time*—
say one or two revolutions.

II. But, more complex in character, more difficult in determination, and more remarkable in
their results, are those irregularities or inequalities of the second class, to which the term SECULAR
—because of the great periods they involve—has
universally been attached. These inequalities
affect the orbits in which the planets move—each
orbit, through effect of the actions of the other orbs
on the planet to which it belongs, slowly passing
through modifications, which sometimes occupy
centuries in their course. The relation of the
two sets of perturbations to each other, has been
represented by the following diagram :—

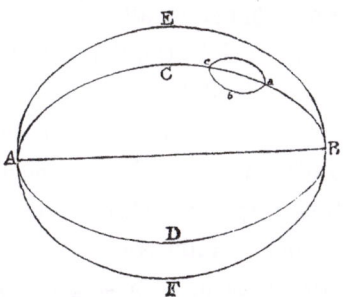

Suppose A D B C the ellipse in which the body
moves at any given period; then, as ages pass

onwards—by the slowest and most gradual evolu-
tion—that orbit will modify itself and change into
A F B E;—this is the change indicated by the SE-
CULAR INEQUALITIES. All the while, however, the
planet is not moving on the surface of the ellipse,
but in a small orbit, a c b, whose centre performs
the revolution in the main orbit; and the distance
to which the planet is carried from that main
orbit by its motion in a c b, is the PERIODICAL
INEQUALITY. It will be readily discerned that the
mechanism of the system, properly so called, consists in the relationship of these Secular Inequalities; for although perturbations of the periodic
class may affect considerably a planet's place, it
is those others which interfere with the arrangements of the system, which show how essentially
each single part hangs on all the rest, and which
alone could affect its stability. Indeed it is in
treating the details of this portion of the subject
that Physical Astronomy has manifested its
greatest power, and where it has most discerned
the perfection of our Planetary Scheme. Let us
rest for a moment, in illustration, on the relations
of Jupiter and Saturn. These great orbs act on
each other variously—producing, as my readers
will expect, perturbations of the periodic class;

but there is one great secular Inequality, than which none within our System indicates more distinctly the delicacy of its relations. It is an Inequality, in the course of whose evolutions the orbits of the two planets present a case of the most exquisite mutual balancing—almost as if one were shifting two balls on the opposite arms of a lever, so that the lever retain its stability.

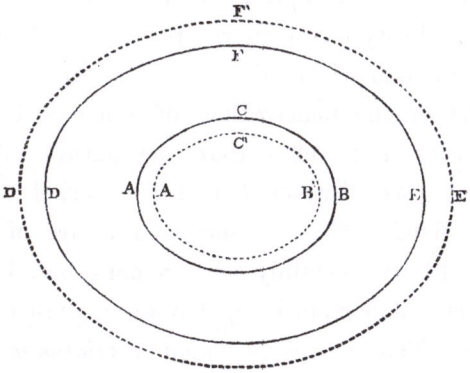

Let A B C represent the normal or mean orbit of Jupiter, and D E F that of Saturn; then, by the action of the planets on each other in the course of centuries, the path of Jupiter contracts itself, slowly creeps in to an inferior limit, A B C′; and during that same time Saturn's orbit as slowly expands outwards, until its limit, D′ E F, is also attained : at this point, the actions of the orbs seem to be reversed—the orbit which expanded before, contracting now, and the other likewise undergoing the inverse change; so that this ex-- quisite adjustment continues—for ever changing, yet for ever stable! The period of these changes is 929 years. It became known, by the comparison of ancient observations with modern ones, that the orbit of Jupiter must have been widening as his velocity was being retarded,—and HALLEY fixed the numerical value of the Inequality: but it was by the penetration of science—by the genius of LAPLACE — that the nature of the change was discovered, and the period of its cycle fixed. The existence and nature of this inequality is certainly very remarkable; but it is chiefly important here, that we apprehend *its origin.* That lies in a singular *relationship existing between the mean motions of the two planets—* the fact, viz., that *five times the mean motion of Saturn is very nearly equal to twice the mean motion of Jupiter:* nor is the result peculiar to these planets: wherever any such relation,—however different numerically, exists between the motions of any two planets affecting each other,—it will obtain a definite expression in perturbations, in every respect manifest, and determinable by observation, like the one whose character I have now explained. How discriminating, then, as well as powerful, the glance we now can cast among the complexities of our system! Resuming our previous hypothesis of an invisible Jupiter, have we not here a distinct means, by which, from the information given by observation, we might reach the most important characteristics of its motions? As the balance of Saturn in those remarkable oscillations, we would require not a new Planet merely, but one whose motions bear towards those of Saturn a *certain definite relation;* and that relation would establish all the important facts regarding Jupiter's orbit! It is not easy to limit the range of these ideas. They point to a knowledge of our planetary scheme, of a kind altogether different from that which we can owe to the telescope—a knowledge of it as it existed prior to its real or visible being— even as it was in the thought of the Great Architect before his creative fiat went forth!

III. The relationships of the separate Orbs, evolving facts so remarkable, my Readers will feel no surprise that the contemplation of the

entire system introduces them to adjustments
equally systematic and unexpected. Our present objects require me to refer only to one,
but that one is sufficiently comprehensive. Not
only has theory shown how every orb is united
to every other by distinct principles or Laws in
the order of our System; but we have arisen to
a few grand propositions involving the whole
mechanism of that System; and which in due
time will afford the fullest command over its
relations with external bodies whose influence is
now unknown. The chief of these propositions,
and the principle on which it depends, are
equally simple. Suppose two bodies revolving
in orbits regulated by *their mutual actions;* lay
down any plane—say the surface of a table,
passing through the system; and having fixed
any point in that table or plane, suppose lines
drawn from that point to every point in the
moving bodies,—this important truth may be
stated; viz., *the sum of all the spaces passed over* on
the table, by these lines drawn from the assumed
Centre to all the moving points, *will, in the* SAME
TIME *be always the* SAME. The first sight of this
grand principle was gained by JOHN KEPLER, and is
expressed in his second Law, regarding the *Radius*

Vector—that Law, viz., which states that if a planet moves round the Sun in an ellipse, the spaces described in equal times by the radius vector are equal. KEPLER derived this law from observation alone—from the facts chiefly regarding MARS, that were accumulated by TYCHO ; but the theory of the subject has extended it to a principle involving all the Planetary System. Assume any plane to pass through that system ; take any point in that plane as a centre; and the areas described by lines from that point to *all* the bodies in motion, will be proportioned to the *times* occupied in their motions. This will take place, *whatever the plane assumed:* but for convenience, as well as more important objects, Astronomers have agreed on a certain definite plane. The characteristic of that plane is this—that the areas described on it, by lines joining the different moving points, with a fixed centre, are *greater* than they would be, if reckoned on any other plane. I am fully aware that the conceptions involved here are not easily realized ; but the value and interest of the Truth will compensate for any effort. Let my reader, for the sake of illustration, assume the Table before him as passing through the solar system, the Sun being in

c

its centre, and Planets moving around him near its surface; then having fixed a central point, it will be observed that the lines from that point to the different moving masses would sweep over various spaces on the surface of the Table. Now, if in this especial plane these spaces are greater than the corresponding ones on any other plane, on this ground it is set apart and considered the EQUATOR-PLANE OF OUR PLANETARY SYSTEM: it becomes that grand plane to which all changes are referred—the plane in itself being *absolutely fixed*—just as the Equator in the Heavens is taken as a fixed circle from which we reckon the position of the individual stars. There is, however, no positive stability predicable of any of the forms or phenomena of Space or Time: and even as our terrestrial Equator is subject to alterations, and has by its mutability unfolded some of the more evanescent changes to which the Earth is subjected; so this grand Equator of our System will, as ages roll onwards, doubtless evolve internal and external actions which escape us now, by *altering its position and exhibiting oscillations amid Space.* If, indeed, there is aught whatever which our scheme contains that is now unknown to us, or if its internal arrangements are subject to influence

by external bodies—by the multitude of Stars
through which it is careering,—these will all be
manifested by a shifting of this PLANE:—it is the
centre, so to speak, to which every vibration of
our nice and harmonious scheme converges; and
at which we shall learn the soonest, not the fact
of the existence of such disturbance merely, but
its origin and entire nature. I believe that a
time will come—although still it is far distant
—when, through the means indicated, we shall
have laid before us the condition and every palpable relation of every constituent of our planetary system, as clearly as the botanist can now
unwind the mechanism of a leaf by his microscope:
nay, farther, and far higher,—building on these
long and incomplete, but, withal, proud labours
of ours—shall we not at some time, pass with
vision as clear among those grand stellar systems,
and having unravelled their order, reach even
their great EQUATOR-PLANE—that normal chart,
stretching through all space, to which the motions
of the entire starry universe report themselves,
and where they impress their harmonies for the
inspection of Man!

Let us resume this last section, before we close

the first part of our work. It leads to three conclusions of high import, and which the reader cannot now fail to realize.

I. These *periodical inequalities*, or disturbances of short period, which cause a body to be either before or behind its otherwise natural or expected place in its orbit,—depending on the *directions* in which foreign bodies lie in regard to it; we would, in case of our former hypothetical search for an invisible Jupiter, be able to conclude only in regard of its *directions* from Saturn at different periods, *so long as our attention was confined to this special class of disturbances* undergone by Saturn. These, however, might amply suffice to guide our telescopes to that portion of the sky wherein, for the time, the new planet would be found.

II. On the discovery of Jupiter by the preceding means, the actual orbit of the star would speedily be fixed by consideration of its observed facts; but in event of the planet continuing invisible, either through effect of distance, or because it reflected no light, the entire and more minute relations of that orbit with the orbit of Saturn—the commensurability, real or approximate, of the periods of the two bodies, &c.— would be unfolded in the course of time, as the vibrations of the orbit of the known planet, or its unexplored secular inequalities, were extracted from the tables of the Observer. No uncertainty could remain therefore;—only Durations must pass.

III. These same results would likewise have issued from the observed instability of the great EQUATOR-PLANE; but I have chosen rather to suppose it reserved for still higher inquiries. The capacities of this mode of contemplating the organism of our system, are larger—much more comprehensive than any other existing process; which, comparatively, are tentative, and suited to each peculiar case. It might have been, for instance, that the irregular and unexplained movements of Saturn, in the case we have chosen to erect in our fancy, were not owing to the action of a planet, but only the most noted manifestations of a disorder affecting all the system, and produced by some remote and general cause. It might have been, that as the Sun careered through

space, in that immense course, by which he
is bound to the great scheme of the stars, influences were shed towards him which told un-
equally on his array of attendants; and these,
whether variable or permanent—instantaneous
or continued — rapid, or of slow growth and
long duration—would inevitably be expiscated,
through their effects in altering the position of
the EQUATOR-PLANE. Does my assertion appear
incredible, or seem to over-rate the power and
penetration of science? Revert, then, to the times
of BRADLEY; and discern him, by the exercise of
his consummate skill, drawing out of slight variations, or oscillations in the equatorial circle
among the stars, motions to which our Earth is
subject, so small and obscure that they had lain
concealed from all the olden times, but which,
notwithstanding their minuteness, have, in these
recent days, enabled another Inquirer, of knowledge and corresponding energy, to penetrate the
internal constitution of our globe! The unity of
this great Universe is unbroken; it is the ap-
pointed theatre of uninterrupted law: and the
power to follow that law through the Heavens,
and to discern by its aid the inter-dependence
of their most varied and gorgeous phenomena,

was the legacy bequeathed to human genius
and perseverance by him who, in this sense,
has never yet been approached—our own immortal NEWTON.

PART SECOND.

THE DISCOVERY OF NEPTUNE.

I.

THE researches of Physical Astronomy, conducted with regard to the problems I have attempted to describe, were rapidly perfecting our knowledge of the Planetary System, when, in the year 1781, the illustrious Herschel altered our conceptions of its extent, by the accidental discovery of the orb that for some time bore his name; but which is now universally recognised, by the more congruous appellation of URANUS. If our acquaintance with the various motions of Jupiter and Saturn—the two planets chiefly within reach of the influence of Uranus—had been as perfect as it is now, the existence of this superior orb would certainly have been suspected prior to its dis-

covery by Observation; but Art had not then
enriched us with those splendid instruments
which in modern times are recording quantities all but evanescent; and the discovery, therefore, came to the scientific world with the fresh-
ness of a surprise. The existence of those connections, however—although they had escaped
our previous observation—was not in the least
degree problematical; and accordingly the task
of modifying our previous conceptions of the
mechanism of the system, so that Uranus should
enter into it as an essential element—present,
not merely in its observed place, but, by its influences through the whole system—was placed
before Astronomers for solution, immediately on
the discovery of the existence of that Orb.
Now this problem evidently concerned,—*first*, the
question of the precise path of Uranus—the
real curve it describes in its revolution around the
Sun : and there are two very essential points as
as to this matter, requiring peculiar notice.

I. The problem itself is this: *From a number
of observed places of Uranus, to determine its entire
path.* It is not necessary to follow any planet
through its *entire course*, in order that we lay

down the path it regularly pursues. Observation so complete, would be needful only if we had no conception of the *kind* of path in which a Planet must be moving; but, since we have a thorough knowledge of the grand LAW that regulates the movements of every orb belonging to our system, a few observations suffice, to enable us to conclude concerning any orb's *entire* habitudes. At the very outset, then, we assume as the groundwork of our Theory of any Planet— and of course with regard to Uranus—*the integrity of the Law of Gravitation in the regions through which that planet moves.* The problem, on the ground of that assumption, is as follows: Reverting to the mode in which every planet must move, as formerly represented by the diagram,

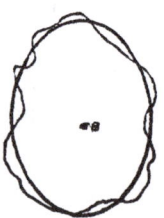

it will be seen that the phenomena of that motion are divisible into two parts—*first,* the normal

Ellipse, depending on the action of the Sun; and, *secondly*—the deviations from it, caused by the disturbing action of the Planets. The two sets of phenomena, however, although distinct, cannot be determined *apart from each other:* it being clear that until we know something of the normal ellipse, we cannot compute the deviations, as these depend on the *distance* of the Planet from *all the other bodies* affecting it: and again, so long as we do not know the amount of the disturbances, we cannot fix how much to take from or to add to, the observed place, in order to arrive at the normal or undisturbed Ellipse. The problem is therefore by no means an easy one: but when our results do not agree with prolonged observation, it is always open to us to try *the effect of an alteration of the normal Ellipse,*—an alteration, however, within limits—viz., the possibility of explaining the deviations from the curve on which we fix, by the action of known perturbing bodies.— The solution, it will be seen, depends, in every view of it, *on our knowledge of the mode of the action of these disturbing bodies.* This knowledge must be clear and definite; or we can reach no solution at all. If any unknown body, for instance, acts on Uranus, then the foregoing principles could never enable us, from a few observations of its place, to determine its true path: and therefore the work of Astronomers must in every such case involve this other supposition—that *we are acquainted with all the bodies that act on the Planet.* Any solution—so far as we have yet traced its progress—thus necessarily involves three uncertainties, viz.:

(1.) *The assumption of the unmodified action of Gravitation in regions so remote from the Sun as Uranus.*

(2.) *The assumption that we have fixed on the true or normal ellipse.*

(3.) *The assumption that all the disturbing bodies are known to us.*

The first definite theoretical step in the inquiry, whose character I have defined, was taken by LAPLACE. His immortal work on CELESTIAL MECHANICS, gave us the formulæ, by which the relative influence upon Uranus, of Saturn and Jupiter, and the Sun, may be computed. He restricted his attention to those planets, because,

from the remoteness and small size of the others,
they alone would have perceptible effect; and his
formulæ therefore comprehended the whole question, as to the action of perturbing influence. I can
scarce avoid, even at this late date, offering my
tribute to this most illustrious Mathematician.
The time is yet fresh within my remembrance,
when, full of the ardour which his name excited
during my youth, I went to Paris, almost with the
sole object of seeing him. Alas for my fond hopes!
his death occurred previous to my reaching that
remarkable city, which thus had become bereft of
what to me was its greatest attraction! Of some
of the most remarkable characteristics of the
Gallic mind, Laplace was an eminent impersonation.
However frivolous in many of their impulsive actions
—often causing, in our more consistent and practical estimation, no slight marvel at their seeming
disregard of the dominion of Reason—no people
have ever shown a finer aptitude for the persevering culture of abstract inquiries. In the use of
the logical faculties, no nation is their superior:
and their impulses are so fresh and powerful, that
no difficulty can thwart or discourage them.
Recently they have evinced, I would say, almost
a *national* aptitude for still higher inquiries. In

DE BIRAN they have produced one who might almost have been a LEIBNITZ; and COUSIN—only by being too much a man of the Forum—seems to have narrowly missed approaching DES-CARTES.

II. It cannot be difficult to apprehend that, in an inquiry like that I have described, the facility of reaching a conclusion, and the certainty of reaching the just one, must largely increase with the extent of that part of the orbit through which Uranus had actually been traced by observation. A very few observations, if they were accurate, and if all the disturbing influences were known, would, in event of our possessing adequate command over Theory, enable us to obtain a right knowledge of the orbit; but should any of these conditions fail, it is clear that the greater the number of the planet's observed places, the sooner would our defects be observed and remedied. Now, M. BOUVARD, who, after 1820, undertook to form tables of Uranus, or, in other words, to deduce the entire orbit of that Planet from observations made during the forty years that had elapsed since its discovery in 1781, found, on due inquiry, that it had been seen at least fifteen times previous to its discovery by Herschel, and recorded in old catalogues between 1690 and 1771, as *a fixed star*. The record of these positions did not, of course, detract in any way from the merit of Herschel's discovery; which was, not that he saw a new star, but that the star in question was a *Planet;* but it clearly very much limited the possibilities of unrecognised error in any theory of the body's motions. For instance, if in any circle, one writes down one set of dots to represent the places of the planet between 1781 and 1820, and another set to indicate those places between 1690 and 1771, it will be readily observed how little remains to be inferred, if we take both sets into account; compared with what would remain to be inferred, if only one set had been established by observation; and it was no wonder, therefore, that BOUVARD welcomed the fact, that the planet had—unconsciously—been so often seen before. On the completion of his calculation, however, he met with a most extraordinary and puzzling circumstance: *The orbit he deduced was found unsatisfactory to either set of observations, and its deviation from the older ones was altogether remarkable;* in other words, he found that the very circumstance which should have enabled him to crown his effort with complete

success, was that which, from some unexpected
cause, rendered success impossible. So long as
he took into account both sets of observations,
or, what is the same thing, so long as he took all
available precautions to avoid error regarding the
habitudes of Uranus—the results of his inquiries
seemed entirely erroneous; in other words, they gave
an orbit inconsistent with the facts on which the
calculations were based. It farther appeared, that
if either set of observations was assumed by itself
as the basis of the orbit required, a result came
out sufficiently concordant with that set, but
wholly discordant with the other set; so that
Bouvard was obliged to conclude that these two
classes of facts were *incompatible:* and the next
point was, what could be the reason of that incompatibility? Perhaps no conclusion was so probable at the time as that to which Bouvard,
amid much misgiving, saw it expedient to come;
viz., that the old observations might have been
less accurate than the modern ones, through
the imperfection of the old instruments; and
that, therefore, they should in this inquiry be
set aside. The conclusion was indeed a bold one,
I had almost written—*audacious:* for among the
observers on whose authority these old places of

Uranus reposed, we find MAYER and the industrious FLAMSTEAD, and, far above all, our incomparable BRADLEY: nor was the ingenious Frenchman blind to the difficulties encompassing the course he found himself constrained to choose. " Such," says he, " is the alternative presented by the formation of the tables of the planet Uranus, that if the ancient observations are compared with the modern ones, the first are passably represented, while the second are not so with the precision they demand; and if either set be rejected, tables are the result which satisfy the ones retained, but do not satisfy the others. It being then necessary," adds Bouvard, " to decide between them, I have held by the modern observations, as being the most likely to be accurate, and I leave Time to come in aid of the difficulty of reconciling these older ones, and of explaining whether it is caused by the inexactness of these old observations, or depends on some foreign and yet unperceived influence to which the planet is subjected." And a very satisfactory light in regard of the difficulty was quickly afforded by " time." If Bouvard had been right, then the movements of Uranus subsequent to 1820, would necessarily have agreed with the orbit deduced from the observations between 1781 and 1820: but, on the very contrary, *the planet has since been moving apparently without the remotest regard to that orbit, and in defiance of all known rule*. Not only, then, was it necessary to reject the old observations in order to sustain the authority of Bouvard's determination, but those more recent ones in regard of whose authenticity and scrupulous correctness, not a shadow of doubt could be intimated, had also to be summarily thrown aside. As with KEPLER, in his pursuit of the true orbit of MARS, no sooner was the Planet, in one part of its course, brought under control and properly enchained, than, lo! at another part he broke from all bondage, and rushed whither he would! I have thought it necessary to represent, in some emphatic diagrams, the actual state of this very puzzling case. In fig. 1, Plate III., if the line A B represent the observed orbit of Uranus, the broken line will exhibit the theoretical orbit,—that calculated from the epoch 1781 to 1840 being traced by the irregular line. Now, the deviation of the two lines—one-tenth of an inch corresponding to one second of space—is the measure of this discrepancy between *Theory* and *Fact;* a discrepancy which in *degree* had no parallel elsewhere

PLATE III.

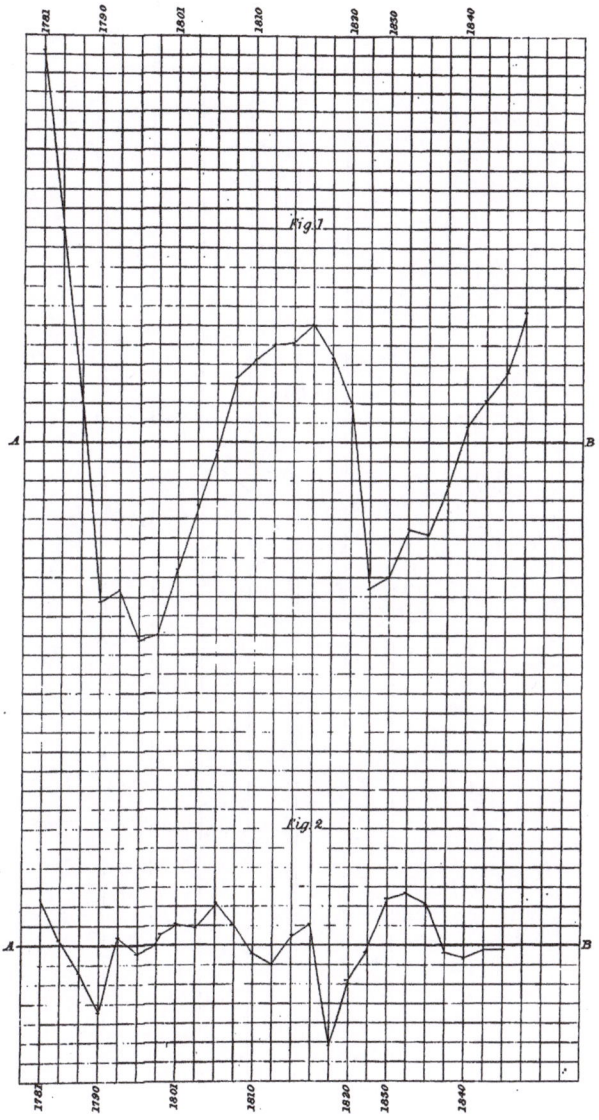

PLATE IV.

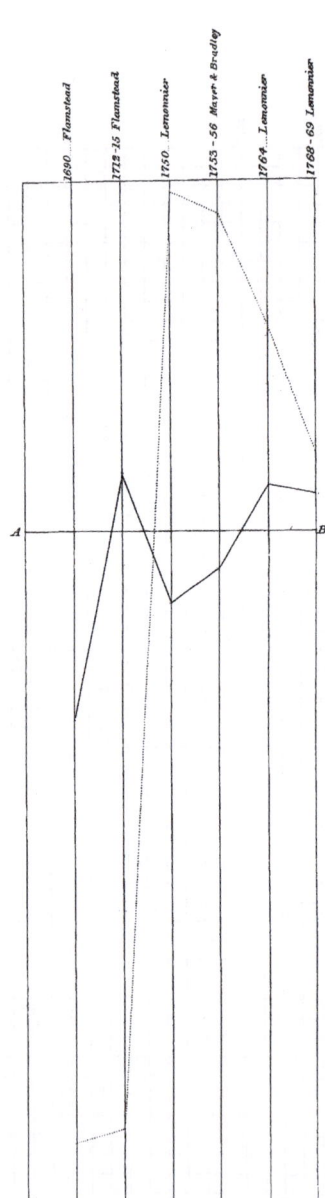

1690. Flamstead
1712-15 Flamstead
1750. Lemonier
1753-56 Mayr & Bradley
1764. Lemonier
1768-69 Lemonier

A   B

in Astronomy. Turn, however, another page, to Plate IV., where the line A B again represents the observed orbit, as determined by those separate acts of observation between 1690 and 1771. The dotted irregular line, on a scale only one-half of the previous one (a second being represented by half a tenth of an inch), is now the representation of the theoretical orbit of Bouvard ! No marvel, indeed, that Uranus had come to be accounted the puzzle of our science,—no wonder that so many minds were turned to this portion of the celestial mechanism, in confident expectation that the anomalies would in time be resolved, by the occurrence of some capital discovery! *

* It is not astonishing that writers of all descriptions, both at home and abroad, were startled by this condition of things, and gave utterance to their feelings respecting it, in various, more or less definite, surmises. I find in a note to a volume of my own, published in 1838, a reference to Uranus thus concluded, "Some great discovery is awaiting us in that quarter." But far more definite conjectures were rife.

II.

How difficult it is to follow, in our Inquiries,
the easy rules laid down by the immortal VERU-
LAM ! Even when the temptations to go aside
are slight, men will not confine themselves to
the only road that can ever lead to a revela-
tion of the secret processes of Nature:—on the
occurrence of any difficulty, they set a-guess-
ing instead ; not after that only legitimate fashion
—the way in which Genius apprehends, combines,
and follows the applicable analogies that lie
around it—but literally almost at random, and
with no apparent aim save the desire to avoid
confessing that there exists something, of which, in
the meantime, they know neither the method nor
the cause. It were positively tedious, were I to enu-
merate the varied fantasies brought forward, not
without adequate solemnity, as probably satisfactory, regarding these remarkable discrepancies;
nor shall I refer, and that very briefly, except to

two. In the first place, the discrepancies were
supposed to result from *accident*,—not to belong
to the *order* of our System, but to originate in
causes of *disorder*; and, according to established custom, alike in Astronomy and Geology,
the action of a COMET was invoked as the
magnus Apollo! One of these wandering bodies,
it was said, had, in the course of its devious path
through our system, come into contact with Uranus—struck it, in fact; and, by introducing a new
cause of motion, produced the discrepancy we discern between the course of the planet at the epoch
of these oldest observations, and its subsequent
movements. It is perfectly certain, indeed, that
such a disturbance would have produced a great
deviation, or apparent breach of continuity. The
planet's path would, in that case, have consisted of
parts of two different orbits, joined at the point of
concussion; but even if its course prior to and after
1780 had been parts of two independent orbits,
internally consistent or harmonious with known law
—*which they are not*—is it not clear that another
comet, and another shock, were requisite after 1820,
to account for the second breach of continuity—
that second departure of the planet from the course
laid down by BOUVARD? If *one* accident might,

without the support of observation of the fact, have been received as possible, the occurrence of *two* such is opposed by all the laws of probability; and the hypothesis might, accordingly, have been rejected, even without the elaborate demonstration of its inapplicability to the details of either case. Nay, this consideration goes further, it forbids our referring—unless distinctly upheld by observation—the anomaly to any accident whatever; for it is quite inadmissible to suppose the *recurrence*, in circumstances so similar, of what in its very essence is *capricious*—an exception, by hypothesis, to all known law or order. We must dismiss, therefore, the hope of receiving a solution from *accident.*—Another assumption which received high favour, must also be mentioned here, as of the kind I term preliminary. The calculations of Bouvard, it will be noticed (see page 45), rest wholly on the supposition of the integrity of the law of Gravitation: now, may not these resulting inconsistencies *have their root in the erroneousness of this fundamental assumption?* The law of Gravity extends as far as Saturn: but, may it not have undergone modification, in those profounder remotenesses from the Sun, through which Uranus revolves in its circuit? This view seemed at one

time a favourite with the illustrious BESSEL ; nor
has it, on the occurrence of other difficulties, been
deemed illegitimate to consider it possible that
Gravitation *may* alter. In truth, we have no
reason to suppose this great Law, as now revealed, to be the ultimate or simplest, and therefore the universal and all-comprehensive form,
of a great Ordinance. The mode in which its
intensity diminishes with the element of distance, has not the aspect of an *ultimate Principle;*
which always assumes the simplicity and self-
evidence of those axioms that constitute the
basis of Geometry : but there is a rule in philosophy, applicable to this matter, that admits of
no dispute. Allowing fully that very few Laws, as
discerned by Man, can be deemed essentially universal—none, it may be, except those first principles of the science of Form, by which so many
of the relations of different parts of the Universe
are determined;—it is yet clear that we are never
entitled to challenge the universality of Laws that,
within our experience, have nowhere failed—*until
every other mode of overcoming the difficulty has
proved of no avail.* If the law of Gravity must be
challenged, then, the time for that is not at the
beginning of our consideration of this difficulty;

but *after* we have tried every circumstance, which —the law remaining entire—might affect the manner of its working, and so have demonstrated that what is now an apparent contradiction, may not be only one of its more recondite and least obvious results.—We must postpone, therefore, that Inquiry, although it had the merit of receiving the approbation of BESSEL.

III.

DISMISSING the idea, then, that such anomalies can arise from accident, and postponing the challenge of the Law of Gravity until every other resource has been exhausted—we find ourselves between the following alternatives: Either we are ignorant of all the conditions within which Uranus is moving,—*i.e.*, we know not the whole of the bodies acting on that planet; or we have not interpreted aright the effect of known conditions. In the course of a truly logical treatment of the question, the latter alternative must first be examined: and this examination necessarily consists in a revision of the theory given by Laplace concerning the action of Saturn and Jupiter,— in the scrutiny of Bouvard's calculations,—and in the discussion of the Inquiry alluded to in page 46, viz., whether, by varying the normal Ellipse within the limits which confine it, and thereby altering the quantities which we ascribe to the deflecting influences of those two large Planets—the existing inconsistencies may not be made to disappear? It was indeed apparently a hopeless idea, that the skill of the author of the MECANIQUE CELESTE might here have failed, or the persevering sagacity of Bouvard; but in presence of a difficulty so startling—of an anomaly so unmanageable amid the harmonies of our System—no weight of Authority could be allowed to forbid our exploring every chance of error; and accordingly, one of the great Men, concerning whom I am to speak, devoted himself at the outset to the immense calculations necessary to construct again from their very foundations, the theory and tables of Uranus. In the work of both of his predecessors, LEVERRIER found room for modifications; and having effected these, he exhausted all possibilities of variation in regard of the normal ellipse. The difficulties were by this investigation, to some extent, diminished; but the startling anomaly remained in fullest force: so that every chance of mistake being banished, the fact stood forth as absolutely and unquestionably true—that there is *a formal incompatibility between the observed motions of Uranus and the hypothesis that he is acted on only by the Sun and known Planets, according to the Law of Universal Gravitation.*—I must here, once for all, solicit the expression of my readers' admiration in respect of the immensity of the sheerest and most unmitigated toil, undergone in this great research. The final results of Astronomy are usually so dazzling, that the mind of the rapt student easily goes in with the idea that the road leading to them is equally pleasant; a road through gardens—among flowers—and by running brooks! Alas! alas! it is a hard and most weary path—across a moor without a blossom! No Siberian steppe can be more arid, than the sphere of these labours of Leverrier; nay, in every portion of Astronomy the labour of dry calculation has become so overpowering, that the resources of Analysis have of late been perseveringly bestowed, on the discovery of modes of shortening and checking these processes, —as one of the most effective means of increasing our power. I cannot but believe—and my belief is founded on a reflection on the remarkably loose, tentative, and artificial condition of our higher calculus—that some discovery here awaits us—some grasping of a new and broad principle as the basis of a new analytic art—which shall again effect in the region of Science a reformation equivalent to that which ranks among the chief of the honours of DES CARTES: and indeed it will be only then, that our growing and extending apparatus for exploring the contents of the skies will reach its efficiency, as a means of revealing the character of their motions—the comprehensiveness and simplicity of their laws.* At present, however, it is " *hodman's work*" that consumes the time of the Astronomer; and this is one among the many points of interest in the case before us. LEVERRIER was detained amid such calculations during the season of the year in which, as a first approximation had informed him, his Planet was passing rapidly from that position in the skies which alone, for a whole year, would permit of its being seen. But he hurried nothing—shrank from nothing; he bore himself as the high philosopher, to whom *guesses* are nothing, unless verified; and he knew that in questions like these, it was only the compatibility of accurate results that could lead to Truth.

* Enough has already been explained to confirm the opinion of the scientific world, that this great reformation might be completed by the most original Analyist of the time, viz., SIR WILLIAM HAMILTON of Dublin. He has begun the work, and shown his eminent power :—let me subjoin my earnest hope, that the character and *rationale* of his comprehensive methods, or rather of his new Analytic Science, may soon be placed within reach of us all—as learners.

IV.

As we advance, we get rid of uncertainties. The problem is now reduced to a simpler form; viz., *Can the anomaly be explained by the supposed action of a foreign and hitherto unknown body on Uranus?* Here, too, however, is choice of hypotheses. Is that foreign body a *new planet*, or is it a body connected with Uranus—a *satellite?* The latter idea was favoured by some Astronomers; but on grounds worthy of no more attention than previous conjectures respecting the Comet. If Uranus was so disturbed by a satellite, that satellite must have been large, and therefore we ought to have seen it even at that remoteness; but what is of conclusive importance, the description of perturbations which alone could have been produced by a satellite, is not the one with which we are at present concerned.

Those would be essentially oscillatory,—the satellite being now on one side of Uranus, and now on the other; and perturbations of that kind are not sufficient for the phenomenon before us. In so far as analogy goes, then, we are forced on the conception that the disturbing cause may be a Planet—yet unseen by the Telescope. This idea, I am perfectly aware, was not a novel one. In a world so busy as ours, and where knowledge, positive and fanciful, belief and idle conjecture, are mixed up and whirled together amid our common speech, in grotesque but inseparable union; it would indeed have been wonderful if this most natural assumption had been overlooked: but rather than seek amidst so strange a chaos for the origin and indication of a grand verity, let us give the credit to the true Workmen, to whose minds the conception was doubtless first recommended by its general verisimilitude, but who were moved thereby, only to work the thought through all its relations, to inquire—in due lealty to Nature—whether this conception was really consistent with her arrangements, and could, when fully carried out, explain the difficulty in its details? "THAT MAN ALONE DISCOVERS WHO PROVES." It is time the truth were apprehended, that there is no royal or easy road to a name in Philosophy.*

The Inquiry, then, points to a NEW PLANET. That word, often carelessly pronounced, became, in the hands of ADAMS and LEVERRIER alone— a definite and practical thought.

If a New Planet—WHERE? By the general order of our System, certain *a priori* grounds of approximation were afforded, which could hardly be much in error;—at all events, they greatly simplified the earliest calculations. *First*, it is a law among the known planets, that they lie nearly on the plane of the Sun's equator. Suppose, for illustration, the Sun represented by a ball laid on the middle of a table, then all the Planets

---

* I cannot avoid subjoining here, in support of the principle alluded to in the text, the following eloquent words of CONDORCET, in his *Eloge de Linné:* — " Trouverait-on une grande theorie dont les premieres idees, les details, et les preuves appartiennent à un seul homme ? Et n'est-il pas juste d'accorder plutot la gloire d'une deconverte à celui à qui on en doit le developpement et les preuves, à celui qui avec autant de genie a été utile qu'a l'auteur d'une premiere idèe vague, sonvent equivoque, et dans laquelle on n'aperçoit quelquefoisle germe d'une dèconverte que par ce qu'un autre l'a deja dèveloppè ?"

would require, if represented, to be laid nearly
on that table;—no one would be found far above
or below it. The new Planet probably following
that law, we should look for it in the Heavens
only through that zone, whose centre is the
Ecliptic, or the annual apparent path of the Sun;
nay, for conformity's sake, we may suppose that
the planet is *in the Ecliptic*; for its utmost probable deviation from that line would produce no
sensible error. *Secondly*, there is another remarkable law, whose origin is quite unknown, but
whose authority, as far as the limits of the orbit
of Uranus, was undoubted. It is this,—Any
planet—speaking in general terms—is half as far
away from the Sun as the one next in order in
ascending, and twice as remote as the one next
in order in descending. If the distance of the
EARTH, for instance, be represented by 1, the distance of MARS is 2, and that of VENUS $\frac{1}{2}$;—and so
through all the known system. This law is sufficiently general to have been the cause of the
discovery of that singular group of infinitesimal
bodies between Mars and Jupiter—the progres-
sion having there indicated a *hiatus*, to be filled up
by a new body; and that group—resembling the
elements or fragments of a single planet—occupies

almost the requisite orbit. It was no unjustifiable hypothesis, therefore, to extend this law to the planet now sought—especially as other although minor considerations sustained the conclusion regarding the body's probable distance. The question thus became as follows:—

"Is it possible that the inequalities of Uranus are due to the action of a planet situated in the ecliptic, at a mean distance double that of Uranus? And if so, what is the place of the planet, what its mass, what the elements of the orbit through which it moves?"

It appears simple now: but assuredly this was the most daring effort ever undertaken by Man. The Earth is distant from the Sun almost by the enormous space of one hundred millions of miles. Uranus is nineteen times farther off, and this supposed Planet must, by hypothesis, be yet twice as far! Can it be, then, that the thought of Man, as the faithful interpreter of Nature's Laws —sympathizing, as it were, with her universal designs—can walk safely amid profundities so dread, and evolve the necessary complements to our planetary Scheme? Are we bound so closely

E

with the system to which we belong, that not a vibration of it can escape us? Is the human Mind indeed, so indefeasibly co-ordinated, even with the grandest laws of this Universe?

V.

I wish I could follow,—with the hope of being accompanied by my Readers,—the march of Leverrier and Adams, in their unwinding of this memorable problem: but their sagacity having for its essential or only possible garb, the symbols and processes of our most recondite Analysis, it is merely a few vague conceptions that I can expect to give of its character and elevation. In what I do say, I shall abide by the processes of Leverrier, partly because they are more susceptible of representation than the more general developments of Mr. Adams, and also because, as a matter of fact, they led to the first discovery of the actual existence of the body sought for, and therefore first proclaimed to the world this new triumph of Science. —There are a few, although only a few, distinct points, which may be generally apprehended.

I. There is a source of misconception at the

very outset of any consideration of this subject, which must be removed. It will occur to many, that since we have assumed that the planet sought for is moving in a circle around the Sun, whose radius is twice as great as that of the orbit of Uranus, it could be no matter of great difficulty to find its place, inasmuch as we have only to observe the *directions* in which its perturbing influence pulls that body at different times; which directions would be an infallible index to where the planet was in its orbit, and therefore in what region of the Sky it should be sought. But, to accomplish this, it would be necessary to know, in the first place, with entire precision, *what amount of the motions of Uranus is due expressly to that unknown Orb;* and the difficulty just is, in separating these with accuracy from what is due to the Sun, and what is due to Jupiter and Saturn. No one—as I have already said—of these sets of motions can be determined without all the others; the process is not first to set aside what is due to the Sun, next what is due to Jupiter and Saturn, and to take as the perturbations sought, merely the residuum. It is a process, on the contrary, which determines *all at once:* the normal or solar ellipse, for instance, does

not appear, unless at the moment which also reveals the other perturbations: and these must continue unknown, until we have likewise obtained the
elements of the orbit of the new planet. The task
is thus—as every one addicted to such contemplations will observe—one of the greatest complexity.
Observation gives us only the total or *gross* result of the combination of a number of unknown
quantities, viz., the elements of the unknown
ellipse described by the new planet, and its mass,
and the elements of the yet unknown ellipse
described by Uranus: and by a scientific appreciation of the *relations* necessarily existing among
these different quantities, we have to judge of
their relative influence in the production of that
total Result—and thus to eliminate their separate
numerical values. It is not to be gainsaid that
problems more difficult than this, have often been
resolved in the history of our conflicts with the
Mechanism of the Heavens; but in character
it was wholly new—nothing like to it having
been ever formerly resolved or undertaken;
and I think it is one of the many incidents
which call for especial admiration, even in a
history of which every step is fraught with un-
wonted interest — that firm but unassuming

record in his note-book, by the resolute and modest under-graduate of St. John's, to dedicate himself head and heart—so soon as duty permitted the adequate leisure—to the struggle with difficulties he has now so nobly overcome!

II. It cannot require acquaintance with technical analysis to enable one to apprehend the nature of one means towards the solution of this problem, viz., the method, adopted by Leverrier,—of *approximation*. A view of the conditions that must determine a final result of only a rough description, would evidently conduct to *limits* within which both the place and the mass of the planet must lie. After the different quantities involved are assorted according to their relations, it is not difficult to conceive how a view of their mode of arrangement would lead to the conclusion, that—without attempting minuter appreciation—the attributes of this body must be confined within certain distinct numerical expressions; and that—as the two most important attributes—the *place* of the orb at a certain epoch, and its *mass*—might be roughly separated, the one having to do chiefly with the *direction* of the disturbing force, and the other solely with its *magnitude*,—it was no unlikely thing

that a preliminary statement with regard to both
could be made: which—being accepted—would
greatly facilitate further inquiry. Leverrier's
first conclusion was as follows :—

"THERE IS IN THE WHOLE ECLIPTIC ONLY ONE
REGION IN WHICH THE PERTURBING PLANET CAN BE
SUPPOSED TO BE PLACED, SO THAT IT ACCOUNT
FOR THE IRREGULAR MOVEMENTS OF URANUS. ON
THE FIRST OF JANUARY 1800, ITS MEAN LONGITUDE
MUST HAVE BEEN BETWEEN 243° AND 252°."

My readers will readily appreciate the extent
of new knowledge acquired by this step. By our
first and fundamental assumption, the new planet
had to be sought only in the Ecliptic—in that
one great circle around the Heavens, and not
indifferently, in any place of the dazzling vault.
—The meaning of the foregoing *numbers* is as
follows: To fix or refer easily to the place of a
star in the ecliptic, we suppose that whole circle
divided into three hundred and sixty equal parts,
named degrees; and we begin to reckon these
from a determinate point, viz., the first point
of Aries—passing all round that great circle.
Observe now the limitation previously made; the

place of the unknown planet was confined by it *within the brief space of* NINE DEGREES *in a circle consisting of* 360°.

This important limitation conducted immediately to a much more distinct enunciation of the planet's place. Leverrier soon discerned that the necessity of satisfying all the relations of the unknown quantities would not permit an uncertainty as to the planet's place of even these nine degrees, and having determined the limits of its mass, he reached the following proposition :—

" *That all the observed motions of Uranus could be accounted for by the perturbing action of a planet, the elements of whose orbit were primarily assumed, whose longitude on* 1st *January* 1800 *is* 252°, *and whose eccentricity and the longitude of its perihelion were determined by processes he had just explained.*"

The mass of the planet he had fixed between the mass of Uranus, and one three and a half times as large.

From the previous theorem it followed that on the 1st of January 1847, the heliocentric

longitude of the required orb must be 325°;—an
astonishing approximation, with which most men
would have been satisfied; but it only informed
Leverrier that a farther accuracy might yet be
obtained !

III. On the 31st of August 1846, Leverrier
produced his last great paper to the Institute.
During his former researches, or previous to his
having selected any part of the ecliptic as more
likely than the others to contain the new Planet,
he had confined himself to consideration of a
certain number of the facts ascertained in regard
of Uranus—selected because of their fitness to
yield a first approximation; but to give his work
the utmost precision of which it was susceptible,
he now employed the whole of the nineteen old
observations made between 1690 and 1771, and a
very large number of the two hundred and sixty-
two places found in the records of the observations of Greenwich and Paris between 1781
and 1845 ;—each of these separate facts giving a
distinct *equation of condition*, as it is termed, or a
numerical value of certain combinations of the
unknown quantities, viz., the correction of the
elements of the orbit of Uranus ; and the mass,

and the elements of the orbit, of the Planet sought
for. No fewer than *nine* unknown quantities
were involved in this work; and he reduced his
equations ultimately to the number of *thirty-three.*
The treatment of these, again, involved immense
toil; and it was while undergoing this last labour
that Leverrier had the mortification to observe
that before his calculations could possibly close,
the planet, in its apparent course through the sky,
would have passed for the year from a position
the most favourable for its being seen. His
labours at length were terminated; and he an-
nounced finally to the French Academy the following elements :—

Radius of the orbit,   . 36·154 times that of the Earth.
Period of revolution,   217·387 years.
Mean longitude, } 318·47
Jan. 1, 1847, } . $\frac{1}{9300}$
Mass,   .   .   .

From which, an easy computation showed, that
the true heliocentric longitude on 1st January
1847, must be

326° 32

instead of 325°

as roughly given by his first approximation.

How singular that scene in the Academy! A
young man, not yet at life's prime, speaking un-
falteringly of the necessities of the most august
Forms of Creation—passing onwards where Eye
never was, and placing his finger on that precise
point of Space in which a grand Orb lay concealed;
having been led to its lurking-place by his
appreciation of those vast harmonies, which
stamp the Universe with a consummate perfection! Never was there accomplished a nobler
work, and never work more nobly done! It is
the eminent characteristic of these labours of
LEVERRIER, that at no moment did his faith
ever waver: the majesty of the Enterprise was
equalled by the resolution and confidence of the
Man. He trod those dark spaces as COLUMBUS
bore himself amid the waste Ocean; even when
there was no speck or shadow of aught substan-
tial around the wide Horizon—holding by his
conviction in those grand verities, which are not
the less real because above sense, and pushing
onwards towards his New World!

IV. Let us recall for a moment the nature of
this entire process—the cause of that singular
confidence which upheld ADAMS and LEVERRIER.

The anomaly by which the investigation was suggested and its course shaped, I previously expressed to the eye in Plates III. and IV. The orbit of Uranus, as explained by the theory of the Law of Gravity, differed from the observed orbit in as far as the broken line in those diagrams differs from the straight one; and the success or verisimilitude of any new theory, simply depended on its power to make these discrepancies disappear. The determination of the place and magnitude of the new Planet, resolved itself into this question, viz., By what hypothesis of this kind can the foregoing object be best effected? Now the reply by Leverrier is contained in fig. 2, Plate III.—the difference between the two lines indicating what remains of that old discrepancy. The argument was, that in thus far, confusion had been resolved, and seeming inconsistency reduced within the universal Order. And it is thus alone that we know Truth ;—we accept a theory, simply because it declares how order penetrates through some section of phenomena : and indeed it is the groundwork—the fundamental assumption—of all Philosophy, that RULE and UNITY exist, and that Man's noblest function is to recognise them.

VI.

WE touch on the close of this long and noble endeavour, viz., the actual discovery of the Planet.—There are several points connected with this act, which it is of extreme importance that my readers rightly understand. The discovery of a planet by the telescope can be made only in one of two ways—by the observation of one of those two features which alone distinguish such a body from a fixed star, viz., its possession of a *palpable disc,* or its having *planetary motion.*

I. If the orb is of sufficient magnitude to exhibit a measurable disc to the telescope employed in the research, its discovery is comparatively easy; for the feature in question wholly separates it from the class of the Fixed Stars. These mighty orbs, although of the magnitude of Suns, are seated so profoundly in space, that

to the largest instrument with which we have yet
examined them, they appear with a brilliance
augmented indeed in proportion to the size of the
Telescope, but still only as *points*, severed essentially from those small orbs whose dimensions we
can descry and compare, and which are our com-
panions and neighbours. It was this attribute
which revealed Uranus to Herschel; and Leverrier threw out the idea that the actual mass
of Neptune, and the augmented power of the Instruments that can now be pressed into service of
such a research, favoured the expectation that, by
its possession of a visible disc, and therefore
without any overpowering labour, this new Planet
would be found.—In several parts of this remarkable work, Discovery seems to have been
attended by a propitious chance; and although,
as we shall afterwards see, the grounds of Leverrier's expectations were here fallacious, his prediction of the actual *apparent* size of Neptune
approached surprisingly to the truth. This disc,
however, although definite and measurable, is so
small as almost to be illusory; and it was not by
it that the Planet was discerned.

II. Unaided by any visible disc, the Explorer

has only one other resource:—among the multitudes of small stars in the quarter of the Heavens
where the unknown orb is conceived to lie, he requires to ascertain whether any one has a planetary motion. But this cannot be discerned by a
single inspection. The motion at remotenesses
like those with which we are now being conversant,
must be so slow, that, for the brief time of one
night, or even of several nights, it may be virtually
equivalent to stillness: so that it cannot be detected save in one way, viz., the careful comparison of the state of the Heavens on one night, with
their state on some other night, separated from the
first by a considerable interval. Now, this comparison is not easily accomplished—nay, it involves
great labour; it requires that *an accurate map*
be made of all the small stars in the region of the
sky under scrutiny, *at these two several times;*
and to do this—to map the small stars in any
region of the sky even once, involves a labour so
great—taking the necessary exactitude into account·—that LEVERRIER gladly expected the
desired result from the visibility of a *disc,*
to avoid what he rightly terms — *ce travail
long et penible.* But Fortune was again favourable!

III. For many years a great enterprise has been in the act of being performed by the Academy of Berlin—chiefly through the instigation of the illustrious BESSEL. Convinced of the great importance of the work, especially with regard to such discoveries as this—the Academy undertook the mapping—with all the precision which our modern Instruments render possible— of the small stars along the entire *Zodiac*, or along that belt of the sky, where—from the analogy of the other parts of our system—new planets might be expected to be seen. The labour required to achieve this was enormous; and it was divided among a great number of persons, having requisite instruments. Now, it so happened that the map of that precise region where the new planet was expected, had been completed by Dr. Bremiker; and it was printing, or just printed, at Berlin:—I believe that the Observatory of Berlin had obtained the proof-sheet. The Astronomers of this Institution were thus in a position of power regarding such inquiries, enjoyed by no other Observatory in existence : they had simply to notice Bremiker's Map and then the Sky—observing if there was a discrepancy between the two pictures, that could be accounted for by the planetary motion of some one star: so that—with their renowned sagacity, and the excellence of their Instruments—an inspection of the Heavens on one clear night might accomplish the resolution of this great problem. And thus it even was; the Planet was discovered actually by M. GALLE, on the very evening of the day on which he received the letter of LEVERRIER indicating its place. The following is the announcement of the final solution of this difficulty in the mechanism of the Heavens:—

BERLIN, 25th *September*, 1846.

SIR,—THE PLANET WHOSE POSITION YOU MARKED OUT ACTUALLY EXISTS. ON THE DAY ON WHICH YOUR LETTER REACHED ME, I FOUND A STAR OF THE EIGHTH MAGNITUDE, WHICH WAS NOT RECORDED IN THE EXCELLENT MAP DESIGNED BY DR. BREMIKER, CONTAINING THE TWENTY-FIRST HOUR OF THE COLLECTION PUBLISHED BY THE ROYAL ACADEMY OF BERLIN. THE OBSERVATION OF THE SUCCEEDING DAY SHOWED IT TO BE THE PLANET OF WHICH WE WERE IN QUEST. . . .

J. G. GALLE.

As ascertained by M. Galle, the heliocentric longitude of the body for the epoch of 1st January, 1847, would be

327° 24

The predicted longitude 326 32

as before stated. The difference was, therefore, less than one degree, or only fifty-two minutes !

F

The entire annals of Observation probably do not elsewhere exhibit so extraordinary a verification of any theoretical conjecture adventured on by the human spirit! M. LEVERRIER received the cheering intelligence, after he had concluded his last paper to the Institute on the subject ; and his bearing was too striking and characteristic to allow me to omit reference to it. "*This success*," says he, "*permits us to hope that after thirty or forty years of observation on the new Planet, we may employ it, in its turn, for the discovery of the one following it in its order of distances from the Sun. Thus, at least, we should unhappily soon fall among bodies invisible by reason of their immense distance, but whose orbits might yet be traced in a succession of ages, with the greatest exactness, by the theory of Secular Inequalities.*" Am I indeed overcharging it, in deeming that the attitude of the Inquirer here approaches the SUBLIME? Standing on the summit of a pinnacle to which the loftiest minds had heretofore looked with rather an aspiration than a hope, his first glance is even farther onwards,—his thoughts stretch towards remoter Altitudes still lying cloud-capped; but which may one day be scaled, and the perspective beneath them spread before the triumphant eye of Man!

VII.

IT now remains that I explain the facts ascertained regarding this new Orb, since the period of its discovery. And in their bearing on several important considerations connected with the System of the World, they possess an especial interest.

I. The Planet, it is found, was seen twice by Lalande in May 1795; and this Astronomer so narrowly missed the honour of adding a fresh constituent to the system, that he rejected his observation of May 8, *because it did not correspond with that of May* 10,—thus losing the momentous truth he would immediately have reached, through inadequate *faith* in his observations! This notice, however, has been of great service, enabling us to fix the approximate elements of the new body much more accurately than could have been accomplished by means of our knowledge of the small part of the orbit which has been seen since the period of the discovery. These elements, as predicted, and as subsequently calculated by Mr. Adams, are the following:—

	Prediction.	Calculation.
Mean Distance, .	36·154 . . .	30·2
Per. Time, . . .	217·4 years .	166 years.
Eccentricity, . .	0·1076 . . .	0·0084
Mass, ...	$\frac{1}{9300}$ . . .	$\frac{1}{23000}$

————

The comparison of the foregoing numbers leads to very puzzling and very remarkable inquiries, —the origin of them all being, that the *true* distance is less than the *assumed* by *six hundred millions of miles;* with a corresponding variation of *fifty-one years* in periodic time: and yet that this FALSE PLANET seemed—even for the delicate purposes to which we have been referring—quite to take the place of that TRUE PLANET, which was *not* wrought with, either by Adams or Leverrier! What, then, is the significance of this extraordinary discrepancy, and how can it consist with the foregoing extraordinary precision? Is the mechanism of the Heavens, after all, so loose-jointed, that

differences of this sort have no effect on its action?
May, indeed, anything be there explained by
anything? We have not, certainly, been asked to
tarry long for the assertion, that the labours of
Leverrier and Adams, belong to the category of
Romance, and that the marvellous discovery
which issued from them, resulted, after all, from
the veriest accident; but I think that, on due
consideration, my readers will not be inclined to
receive this as a true account of the case. To
explain the matter properly, however, I must
solicit an earnest and sustained attention.

The curious subject evidently consists of two
parts:—we must explain, *first*, Why a false planet,
affected by so great an error in distance and periodic time, could so far remove those anomalous
inequalities of Uranus, that there remained only
the small residual quantities I have previously
spoken of? and *secondly*, Why, although incum-
bered by these large errors, Adams and Leverrier, could predict so nearly the true Planet's
place? These inquiries to a large extent involve
each other, but for the sake of distinctness I
shall keep them separate, although at the hazard
of repetition.

1. Starting, then, with our first inquiry, let us at the outset get rid of all difficulties connected with the discrepancy of the *mass*. It will be readily seen that the nearer the planet, the less that mass or that intensity of attractive power which would suffice to affect Uranus so largely as the facts indicate; so that the alteration of *distance* in this respect *involved* alteration of *mass*. In so far, then, as the *quantity* or *amount* of the perturbation is concerned, there is no special difficulty,—that being provided for by modification of the mass; but with regard to the *direction* or *kind* of the perturbation, the case is not so simple,—the great difference in the periodic times of the false and true bodies, necessarily causing them, at different epochs to lie in very different directions with regard to Uranus. We shall be greatly aided in this part of the inquiry by Plates V. and VI., which present the relative positions of these bodies and Uranus, during the various periods at which we have obtained *facts* from observation. I have supposed in both, that the orbits are circular, and the relative places are indicated by the dates corresponding to them. Now it will be noticed in Plate V., which refers exclusively to the epoch of the *modern observa-*

PLATE V.

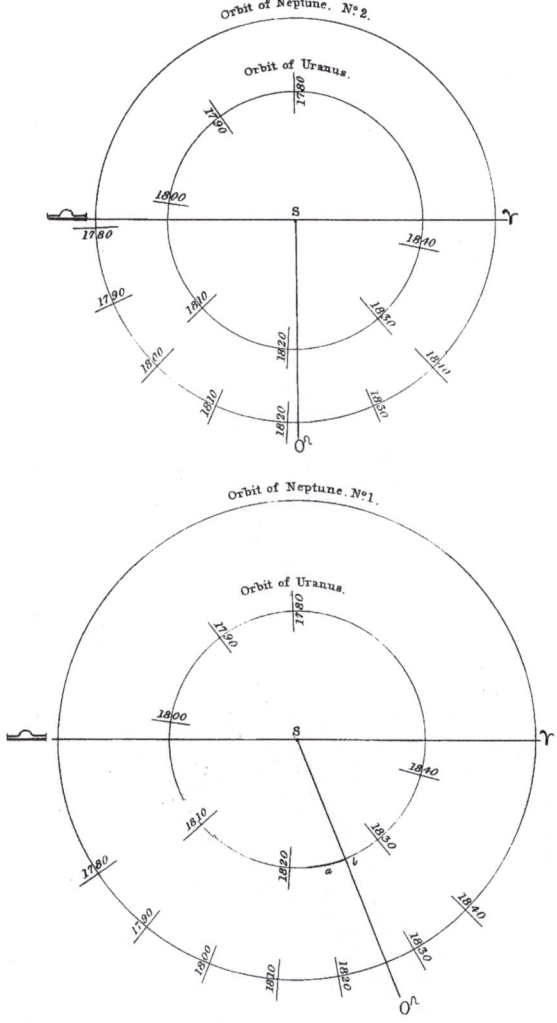

Orbit of Neptune. N.º 2.

Orbit of Uranus.

Orbit of Neptune. N.º 1.

Orbit of Uranus.

tions, that excepting at or near *conjunction*—the point marked by the line drawn from the centre —the *relative positions* of the two planets in respect of Uranus *do not egregiously differ :* throughout the entire course, the *kind* * of the perturbation must be the same in either case, except in the brief part of the orbit, *a b,* marked by a thicker line; during Uranus' motion, in which the bearing of the inequality would in the two cases be *opposite.* This would certainly be the case if the orbit of the planet were *circular,* as I have represented; but it may easily be supposed, that by the adoption of an *elliptical* or *oval,* instead of a *circular* orbit, this cause of discrepancy might be made to disappear, simply by effecting the transfer of the point of conjunction in the case of fig. 1, to a point corresponding with the true epoch, as represented in fig. 2. The adjustment now referred to, was, in fact, one reason for the adoption of that great eccentricity, or of that large deviation from the circular form of orbit, which was found necessary, with the supposition of the false distance;—by which, too, other slight discrepancies of direction were lessened by important

* I mean by similarity of *kind,* that it will either be *accelerative* or *retarding* in both cases.

quantities. The corrected curve in Plate III.,
viz., fig. 2, it must be noticed, by no means evinces
an entire correspondence of theory, as given by
the false planet, with the realities of observation;
and the inequalities still remaining must be
placed to account of the fact, that a perfect or
complete adjustment could not be effected even
by variations in the form of the Ellipse.

The illustration now given, however, goes
back no farther than the epoch of the discovery
of Uranus; we must try it, where explanation is much more difficult, viz., with respect
to the observations between 1690 and 1771.—
To prepare for this portion of the inquiry,
I must recall attention to Plate IV., where the
former bewildering discrepancy, and the recon-
cilement by the false planet of observation with
theory, are pourtrayed. The orbit as amended
by aid of the false planet, is there represented by
the unbroken line—a line which certainly much
more closely corresponds with the observed path,
A B, than the dotted course exhibiting the anomalies as they previously stood; but which —
especially since the discrepancies are measured in
this plate, only on half the scale employed in
Plate III.—by no means shows any marvellous

PLATE VI.

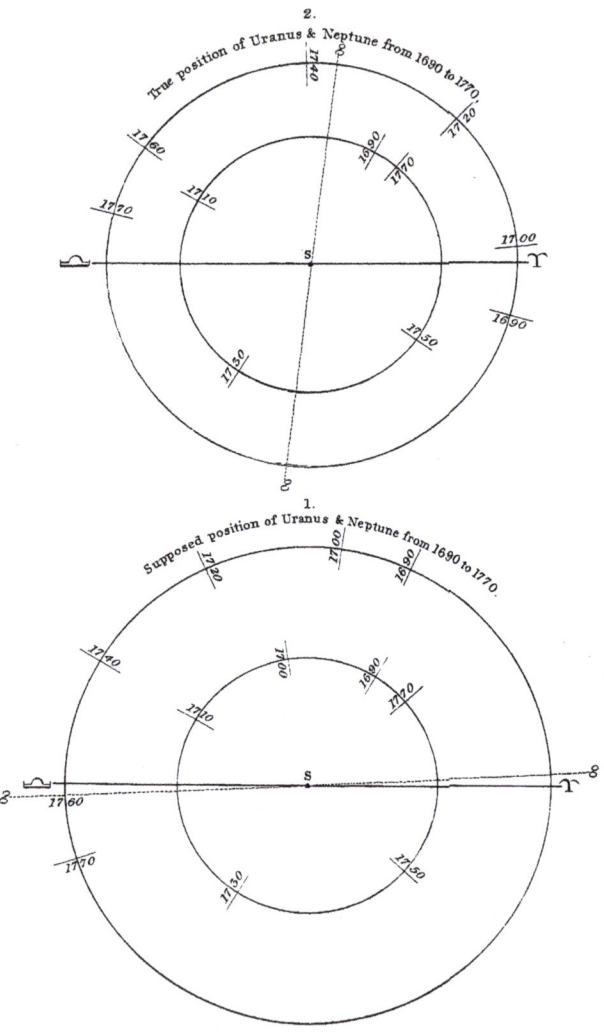

2.

True position of Uranus & Neptune from 1690 to 1770.

1.

Supposed position of Uranus & Neptune from 1690 to 1770.

J. Bower Sc! Edin?.

coincidence, or very successful adjustment. Let it be borne in mind also, as an essential preliminary, that the portion of the line *above* A B, marks where the new theoretical orbit errs by excess, or where the planet's theoretical place is *in advance* of its observed place; while the portion *below* the same normal line manifests that at certain epochs marked at the top of the diagram—the calculated orbit errs by defect—giving places for Uranus *behind* its real or observed places :—A simple inspection of the diagram will make it clear, that the residual errors in *defect*, are considerably greater than the errors of *excess*. — Turn now to Plate VI., which, after the plan of Plate V., exhibits the relations of direction between Uranus and the true and false Planets, at all those epochs between 1690 and 1770, at which observations exist: and as a key to the estimation of the scheme thus represented, observe, by aid of the diagram following, how a planet A, revolving around a centre O, and disturbed by the external body B, must be affected according to its position in regard of B. A, it is sufficiently clear, will be *accelerated* or drawn onwards in its course, during its motion *in the semi-orbit* A" A' A ;—more

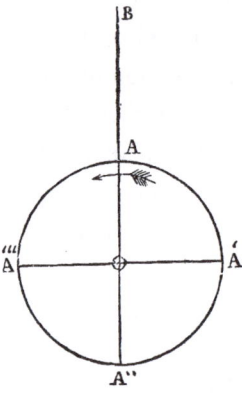

accelerated, however, whilst in the near quadrant A' A than in the quadrant A" A';—and, on the other hand, it must be *retarded when in the semi-orbit of* A A''' A;" and more, during the quadrant A A''' than when in the quadrant A''' A". Minuter, but withal perfectly simple considerations, would bring out other facts of consequence, as to these accelerations and retardations; but what I have just stated will suffice. Let us now examine then, with some care, fig. 1, Plate VI., which shows the relations of the *False Planet* with Uranus during these eighty years. The epoch of 1690 may be termed the epoch of the conunction of the two orbs, or when the energetic retarding action would begin: the action due to the first quadrant, would continue till 1730 ; after which the relation of the bodies would be that of the second quadrant, and would so continue till 1760. These entire effects are as follows:—

ACTION OF THE FALSE PLANET.

From 1690 to 1730.—The powerful retardation of the first quadrant,— conjunction having just passed.

From 1730 to 1760.—Action of the second quadrant—still a retarding, though less powerful, action.

From 1760 to 1770.—Action after opposition— that of acceleration, but comparatively gentle, and as to quantity ineffective.

Let me now trace, by aid of fig. 2, in the same plate,

THE ACTION OF THE TRUE PLANET.

From 1690 to 1700.—The latter part of the first quadrant — the action retarding.

From 1700 to 1737.—Second quadrant to opposition—action retarding.

From 1737 to 1770.—Second semi-orbit—action accelerative, growing in potency; but only approaching the second quadrant.

A single glance will discern effective contrasts in these two schemes. Speaking generally, and without regard to minutiæ, they are these:—

(1.) Immediately after 1690, and for several years, the strong retarding effects of the false planet have no equivalent in *quantity*, among the actions of the true planet.

(2.) Between the years 1737 and 1760, the effects of the two orbs are opposed—that of the false planet retarding; and that of the other being accelerative.*

* These results might, of course, be somewhat changed by

If, then, the diagram in Plate IV., representing the abatement made in the anomalies of Uranus, by the hypothesis of the false planet, be further corrected in accordance with the difference of the action of the false and true disturbing bodies, the projections of the orbital line below A B, will be greatly abated: and farther still, *the epochs of the corrections, correspond with the most anomalous of these projections.*

I trust that it is quite unnecessary to remind my scientific readers, that I am offering only a rude solution of a very difficult case; but I think it will be seen from the previous result, that the substitution of the true planet, must—so far from in· validating the former investigation—*cause the anomalies still remaining almost wholly to disappear :* and it cannot longer continue a puzzle that the assumed action of the false planet enabled us to explain to a large extent—although with a considerable *margin*—the existing anomalies of Uranus.

2. Our second inquiry at first sight seems still

the supposition here also, of suitable elliptic orbits: but I have neglected such modifications in this case as insignificant in quantity.

more difficult. Granting that, as above, the hypothesis of that false planet would in so far abate
those obstinate inequalities of Uranus, is it not
strange that it should lead to the detection of the
place of the true planet to within *one degree?*
Speaking *a priori*, one would have expected from
such an investigation, nothing farther than proof
—arising out of the great abatement of these inequalities—that the mode of explaining them now
adopted was the true one; and so have encouraged
the repetition of the inquiry with different distances until the actual body should be found: so
that it demands explanation still, why a result
so remarkable as that accurate prediction, befell
in this case; or what is the same thing, why the
place of that false planet, as seen from the Sun,
coincided within a hair's-breadth with the place
of the true one? I have already explained how
the theoretic orbit in question could be modified,
so that the body moving in it, according to established laws, might, even through a certain space
occupy about the same *directions* in respect of the
Sun and Uranus, as the true perturbing body
actually does occupy; in fact, one part of the
problem solved by Adams and Leverrier, virtually
was, the discovery of the elements of an ellipse

which, with the false axis, would yet exhibit the
body moving in it, coinciding in direction from
the Sun, with the true body, through as large a
space as possible. But, assuredly, the two orbits
could not be thus brought into accordance, except
through comparatively a small space—on either
side of which, the bodies would cease to coincide;
no possible power or artifice being able to keep in
the same apparent place in the sky, through all
their course, Orbs, one of which completes its circuit sooner than the other by *half a century*. This
power, then, is limited: but it exists; and in accordance with that singular good fortune which has
marked every step in this inquiry, the adjustment
was needed *when the planets were in that part of
their orbits where it could be most easily made, and
its efficacy extended over the longest period!* This
fact—one sufficiently curious—will be readily comprehended. In the first place, it is clear that when
two planets are in the relative position represented
by A and B in the diagram of page 98—it is of no
consequence what is the distance of the disturbing
planet B;—any distance in fact will represent any
effect on Uranus if we retain power to *alter the
mass*. The assumed distance, in such a position,
can have no effect over the *direction* of the dis-

turbing body; as whatever that distance, it will
always be directly *behind* Uranus: and it must
likewise be manifest, that for a considerable time,
alike previous and subsequent to this epoch, error
of distance will have the least possible effect, or
rather much less effect on that direction, than
it could have had at any other conceivable time.
It must be noticed farther, that the duration of
this auspicious period of actual or proximate conjunction, is not to be measured on the supposition
that the one planet is moving swiftly, passing
the other at rest. The motion of Uranus being,
in the direction of that of the perturbing body, the
duration of their proximity would be so far prolonged, that modification of the orbit, or its change
from a circular to an elliptic curve, would suffice
to identify the true and false planets in direction,
or speaking scientifically, in heliocentric longitude,
for the forty years preceding the epoch of conjunction in 1820, and also for the years that since
have succeeded. That identification, however—
that *happy* identification, inasmuch as it led to
the memorable prediction of the place of the
Planet—could not have endured long: the ellipse
was still tentative only—adjusted to a comparatively small portion of the orbit, and every succeed-

ing year would the more strikingly unfold, through the new deviations of Uranus, that the true Planet as yet had not been discovered. Mr. Adams, indeed, saw, and aright interpreted, the monitions of the more recent years; for the differences which during these years still existed between theory and fact, impelled him to the supposition of a diminished mean distance. More consonant results were his reward; and if the Planet had not in the meantime been discovered, the method of our countryman would, in no long interval, have established the orbit as it is. If, however, we had attacked the problem when the Planets occupied different portions of their orbits—for instance, during the period between 1690 and 1771—it would have been wholly chimerical to look for such coincidence. With an erroneous mean distance, no correspondence, in that epoch of their revolutions, would have been possible; and the predicted place of the disturbing body would necessarily have been true, only within wide limits. We owe success, then, in some measure to the auspiciousness of the Time at which the task was undertaken—*task* I say, for in modern times the powers of the Sibyl belong to the laborious men alone; but, nevertheless, the time was so auspicious—the behaviour of Uranus was so like what appeared to belong to a case of conjunction, that its existence ought to have been surmised; and, apart from all intricate calculation, this should have induced us to explore with our Telescopes, a special and not very extensive portion of the sky.

II. It can scarce fail to impress one, that the difficulties over which I have just glanced, arise from the infraction of a Law which had not only prevailed through all the other domains of the Solar System, but in former times been the means of conducting us by prediction even to the knowledge of other orbs. BODE'S Law or expression of the order of the planetary intervals certainly holds as far as the orbit of Uranus; and, therefore, the outward extension of it seemed upheld by a large induction of facts: but it must not be overlooked, that Induction is never safe, so long as it reposes merely on *outward resemblances, or results.* Bode's law is not a principle, but the issue of some principle in the meantime quite unknown; and it was, therefore, impossible to judge *a priori* whether the very principle lying at its foundations, did not demand a modification of its forms, on reaching these ulterior

distances. That it has done so, is now established—a new truth among the elements of our
System : but, as I have said, the Law, as formerly apprehended, did good service in its time.
It was the abnormal blank between Mars and
Jupiter which caused that association of German
Observers, about the close of last century, that
led to the discovery of the group of the small
planets*—a group of exceeding interest; and to
whose number, within the last eighteen months,
three new ones have been added. It is a group
connecting the class of large planetary orbs with
existences seemingly of a different order or grade
—descending even so far as those streams of
meteorites, that once again seem to promise to
revisit our Heavens.†

III. But the new Orb, if thus indicating an infraction of what before we reckoned our system's
established order, appears in other particulars

---

* PIAZZI, however, the first discoverer in this region, did
not belong to the Association.

† This prediction is perhaps not much more than a guess.
If, as I suspect, the nodes of these meteor streams revolve,
I would be inclined to think that we are again approaching
connection with the streams themselves, by the very remarkable display I witnessed in August last.

108

destined, if not to restore, at least to confirm it.
The first light obtained respecting its physical
character, we owe to Mr. LASSELL of Liverpool;
and that light is of great importance. It appears
scarcely within reach of doubt that the orb is
surrounded by a RING like SATURN'S; at least it
will be remarkable if this excellent and accurate
Observer has been mistaken here—although as
yet he has put forth his views only as strong sus-
picions.* But the detection of one satellite by
the same gentleman is now beyond challenge; nor
can there be great error in the period he has
assigned to the revolution of this body, viz., about
six days, in an orbit almost exactly the size of
that of our moon. Here, then, at this vast remoteness, we have simply the repetition of
schemes with which we are already familiar,—the
existence of the *Ring*, should that be confirmed,
connecting such formations still more closely with

* We must not forget, however, that at first, Uranus was
fancied to have a ring, or rather rings; so that the observer,
when examining objects of so great a remoteness, must be
subject to some special delusion in this respect. Doubtless,
Mr. Lassell's caution has been increased by this fact.—The
foregoing facts respecting the *Satellite* have just been confirmed by M. STRUVE of POULKOVA, and are therefore beyond
all doubt. Most heartily do I congratulate Mr. Lassell on
this gratifying success!

the order or normal state of the planetary system;
and in so far as one can yet judge, the position
of the Satellite, and the direction of its motion,
do not partake of those anomalous peculiarities,
which to some extent divorce Uranus from
the conditions within which the other orbs hold
their existence. The motion of the satellite is
direct; and it is not strikingly, or more than
usually, removed from the plane of the Ecliptic.

IV. How extensive are the new adjustments,
how serious the task devolving on the physical
Astronomer, through the introduction of this
formerly unknown Orb, amid the arrangements
of our planetary scheme! It is a mechanism which
resembles an elastic cord ; — touch it in one
part, and the vibration passes onwards like a
wave, until it reach the outermost extremities.
The new planet must disturb Saturn, so that
part of Saturn's inequalities, formerly explained
by other causes, find their solution here. Actions
formerly attributed to other bodies—say to
Jupiter—must now, therefore, be withdrawn from
among their efficiencies; and this will alter our
opinion of their masses or weights,—an alteration that constrains changes in their relations

with the *interior* planets. The entire system, therefore, feels the introduction of the novel inmate; and skill consummate, as well as industry that will not tire, is needful again to adjust and rightly define its harmonies. It is happy indeed for Science, that both the illustrious men to whose genius and energy the world is indebted, for its knowledge of this distant Orb, are already active in pursuit of the objects their own discoveries have created. They are labouring at different parts of the great subject —Mr. Adams, in the meantime, restraining his regards within the environs of his former triumph; and Leverrier beginning, with a view to work outwards, in the immediate neighbourhood of the Sun. All success to them, whether as separate or as fellow-workers! The world will expect what they can easily give it—a new and more perfect MECANIQUE CELESTE.

PART THIRD.

PERSONAL CLAIMS AND DESERVINGS.

IT is painful, that at the conclusion of a discovery which will be ever memorable—at the consummation of an achievement, which will rank through all time, as one of the most daring endeavours and noblest triumphs of the Human Reason,—we find ourselves in the midst of broils and ungenerous jealousies and resentments, of manifestations on the part even of distinguished Intellects, of those mean and disingenuous efforts to undervalue and detract, that constrain one to blush red on mention of Man's unchallenged greatness!—Since so it must be, however, let us look frankly at the question—resolved at once to see everything, and to state the truth without colouring or reserve.

I believe that, in this case, as in all similar ones, frankness and natural honesty are in the end the best for all parties:—a high Intelligence being much more injured by apologies, than by the distinct avowal that for once, error has overtaken it.

I.

THE question of first importance has to do with the relative claims of ADAMS and LEVERRIER; and on this matter there is doubt or hesitation no longer; it has been judged and finally settled by the all but unanimous voice of the scientific world. The cardinal facts are these:—

1. Leverrier's first approximation was laid before the French Institute on 1st June 1846, and his complete deduction of the elements on 31st August of the same year. Up to which time, no other announcement respecting that important subject had been made public.

2. About the close of October 1845, the elements of the same perturbing body were communicated to the Astronomer Royal at Greenwich —agreeing with the place determined by Le-

verrier within one degree; and in the early
part of September of the same year, these elements had been put in possession of Professor
Challis of the Observatory of Cambridge. The
reality of this important fact is as susceptible of
being established as that of any fact that occurred
yesterday; nor is it now questioned. The full
investigations by both these excellent inquirers,
have now also been for a considerable period before the world; and their several textures are
such, that no light, direct or fortuitous, can be
supposed to have passed from the one to the
other:—the two, therefore, stand forth independent; they are PEERS, The great problem was
resolved by both with consummate success; and
every one saw with unconcealed pleasure, at their
recent and first personal meeting at Oxford, how
thoroughly the sympathies of companionship ex-
cluded all that was ignoble from their rivalry!—
Gladly, too, would I now say—*Content;* for with
one exception all the Astronomers of Europe
have done justice to the merits of both, as men
who have accomplished, and, from whom, in fullest
confidence, they anticipate so much: but there
are two facts which must be noticed in barest
justice to Mr. Adams.

First, Let me deplore the conduct of M. ARAGO. Less for the sake of Mr. Adams, than of one who ought to have been generous, whose distinguished place in the Republic of Letters, surrounds him with responsibilities more grave even, than the duties he owes to France,—a man with whom every other man of genius, be his birth-place what it may, is entitled to challenge a common brotherhood,—assuredly for the honour of M. Arago, must I lament the following explosion of Gallic intolerance, which I fear for ever incapacitates him for the office of judge:—" M. Adams n'a pas imprime, meme aujourd'hui, une seule ligne de ses recherches; il ne les a communiqué a aucun societe savante: *M. Adams n'a donc pas le minute titre valable pour figurer dans l'histoire de la decouverte de la nouvelle planete.* . . . . . Eh bien! il est facile de *prouver* que M. Adams n'a jamais resolu le problem qu'il s'etait proposé. . . . . . M. Adams *n'a le droit de figurer, dans l'histoire de la decouverte de le planete* LEVERRIER, *ni par citation detaille, ni meme par le plus legere allusion.* . . . . . Aux yeux de tout homme impartiel: cette decouverte restera un des plus magnifiques triumphes des theories astronomiques, une des gloires de l'Academie, un des plus beaux titres de notre pays a le reconnaissance et a l'admiration de la posterite!"

M. ARAGO, fatally for himself, forgot that the power of GENIUS is an auspicious one: it is protective of the weak, like the ancient chivalry : the power to perpetrate and perpetuate injustice belongs to the SWORD.—The decree of the Secretary of the Academy was hurriedly reversed; and in accomplishing the act, the clear-sighted and earnest Germans spoke the earliest:—they asserted the rights of Mr. Adams, as history will confirm them ; and named the planet, NEPTUNE.*

But, *secondly*, what seems even a graver wrong still remains unredressed. It is generally known that on the announcement of this remarkable discovery, the Royal Society of London decreed to M. Leverrier the loftiest honour they have the power to bestow; and that the Royal Astronomical Society, after much and repeated deliberation, declined any adjudication in the case. Now, with regard to that latter body, pro-

* It will be seen from the foregoing extract, that Arago, in the name of the Academy, adopted the designation of LEVERRIER,—a name which, as compromising the position of Mr. Adams, was at once rejected by Gauss, Encke, and Struve.

vided it has no power to give two medals, and
also—what is very essential—that it has no right
to regard itself as a *National* institution, having
duties to perform in respect of British science, as
well as the privilege of manifesting its sympathy
with successful effort wherever originating,—on
the ground of these two assumptions, one may
not quarrel with it, but only regret that a false
constitution, upheld, it may be, by falser delicacy,
prevented a fitting laurel being placed by it on
the brow of a young Englishman, in a case where
none worthier could be found. But, admitting,
though with undisguised reluctance, this especial
and unfortunate excuse, what shall be pled in be-
half of the other association—the most venerable
through its history, in Europe—which in its place,
as the assumed and sustained head and fosterer of
British endeavour, could—in that hour, too, of
difficulty, it might even be of despondency,
because of misfortune and menaced injustice—
overlook Mr. Adams' ever-memorable efforts, and
without one word of apology or explanation—
without inquiry even—rush past to his rival, and
lay its honours at his feet ! I can comprehend
the vindictiveness of Arago. I can understand
on his part a vehemence, prepared even to crush

—worthily or otherwise—alike truth and opposition; for *La Grande Nation* has never been in high repute for tenderness towards men who would share its glory, or of toleration for foreign claims.* But that the body in England, whose duty—if not its foremost one, certainly with lower sanctions than none—has been understood by the State to be protective of the rights of its meritorious countrymen—that this body, composed of the elite of British men of science, and of British noblemen and gentlemen, should have bowed itself before Gallic pretension, and purchased some hollow compliments about liberality and freedom from prejudice, by the sacrifice of the then obscure graduate of Cambridge; yes! this almost inclines one to the hope that the name of Adams may never illustrate its roll! Of this act of our Royal Society, I have heard but one opinion; and it is farther unfortunate, that without such a retracing of steps as false pride is always sufficiently potent to prohibit, the error and injustice do not appear remediable.†

* How often French public acts and disquisitions remind one of the ancient—" We and the barbarians."

† If there is any difficulty in the question between Adams and Leverrier, it must arise solely from one not observing on which special point judgment is required. If, for instance,

One thing, however, is remediable, viz., those faults in the constitution of this important Society which render so great and grievous erring, possible at any moment. The occasion ought not to pass, without serious inquiry as to the nature of that malformation which has thus far denationalized it, or so lamentably weakened its sympathies with merit, however illustrious, provided that at the time it is untitled and obscure : and also wherein is that difference which in the Institute of France has substituted for cold hauteur—uninformed as freezing—an active and earnest friendship for any Frenchman, whatever his condition or existing fame, who manifests desire and ability to illustrate and enlarge the glory of France ?

it is asked through whose means the discovery of Neptune was first declared ? Doubtless it was by Leverrier's. But if the question of merit is separated from accompaniments of good and ill fortune—if it is asked who accomplished the great feat—the feat distinguishing either Inquirer,—the answer is just as clear and ready—It was done by both, and chronologically (although this is of no moment in the question of honour) first by Mr. Adams. There can be neither doubt nor hesitation on either of these matters.

II.

THE most difficult part of this unfortunate subject remains, viz., Why those prior researches by Mr. Adams were not made known to the world when completed, and why they were not acted on ?* As I have already hinted, I shall speak of this portion of the history with entire frankness;

---

* M. Arago, of course, insists heavily on the non-publication of Mr. Adams' researches : but the logic of the Perpetual Secretary is here grievously at fault. An act of publication, indeed, settles the point of priority in discovery in a way that is absolute and undeniable; but it surely does not follow that all other evidence is thereby rendered useless ! A fact may be established in many ways, although one way is the best; but the absence of the means of proving a point in that best way, cannot deprive a man of his right to prove the same point otherwise ? The question rather is, whether this other evidence is sufficient : and it would be a singular court that could doubt the evidence on which the reality of Mr. Adams' pretensions rest. M. Arago should rather have been glad of that fortuitous non-publication; for it alone left the way open to Leverrier.

but whilst I cannot pay any of the celebrated persons whose names have been mixed up with these events, the false compliment of deeming that their usefulness and high repute, cannot bear their being convicted of a mistake, I repel with something more than earnestness, the desire to dwell with undue pertinacity on any error fortuitously committed by men whose services towards science have been, and are illustrious—far more the inclination to exaggerate such an error, so that discontent and mediocrity be consoled. The case, as I apprehend it, stands thus :—

I. The facts, or rather the results of his inquiries—that is, the place of the new planet at a specified time, and the residual errors—were communicated by Mr. Adams, alike to Professor CHALLIS, of Cambridge University, and Mr. AIRY, Royal Astronomer at Greenwich, during the autumn of 1845:—why, then, did these eminent men not act on a prediction so memorable, or at least examine the process leading to it? Are they indeed not chargeable with breach of duty to their high positions—with culpable negligence regarding the advancement of science? Now, keeping steadily in view, that a course of conduct with unfortunate

H

results, is easily confounded—not in the popular
mind merely, but in all minds smarting under dis-
appointment—with conduct morally wrong, there
are several considerations for which, with the earnestness of good faith, I solicit a conscientious appreciation. And in the first place, I believe it would
be difficult to find two men in Britain, or in any
country beyond it, whose time is more thoroughly
devoted, at every available instant, than is the
case with Mr. Challis and Mr. Airy. The labours
undergone by the former observer—in which
he has by far too stinted aid—are established
by his annual volume of results; and the Observatory of Greenwich, under care of the present
illustrious Astronomer Royal, is still—notwithstanding the great efforts at Poulkova—the Model
Observatory of the World. Not for arrangement and aim merely, is it thus distinguished, but
for the extent and efficacy of its work: nor is
it without a degree of national pride that I here
record a recent tribute by PONTECOULANT—the
worthy successor of Laplace, in the Science of
France—a man who is too faithful, too loyal to
Knowledge, to repel or conceal the pleasure afforded him by this noble Institution:—

"Dans un voyage que j'ai fait dernièrement à Londres,

je n'ai rien eu de plus pressé que de me rendre à l'Observatoire de Greenwich, et j'ai dû à l'obligeance de M.
Airy, aujourd'hui directeur de cet observatoire, le plaisir
de le visiter dans tous ses détails. Sans doute le nom-
bre et la beauté des instruments est ce qui m'a frappé
d'abord ; mais ce qui m'a à la fois étonné et charmé
davantage, car nulle part je n'avais rien vu de pareil,
c'est la régularité avec laquelle se font les observations et
l'ordre qui préside aux travaux de toutes les personnes
attachées à cet établissement. Tous les astronomes-ad-
joints travaillent dans un appartement commun et sous
les yeux de leur habile directeur, en sorte que si une dif-
ficulté surgit elle est aussitôt aplanie. On sait que M.
Airy a introduit à Greenwich, comme il l'avait fait à
Cambridge, l'usage d'observations fréquentes des planètes
trop négligées dans d'autres observatoires ; chaque jour
les observations faites la veille sont réduites et comparées
aux résultats de nos meilleures tables, ce qui épargnera
un long et pènible travail à ceux qui voudront employer
ces observations pour construire des tables plus parfaites ;
enfin, pour plus de régularité, des tableaux imprimés sont
mis à la disposition de chaque observateur, qui n'a plus
qu'à en remplir les blancs, d'après les résultats de ses
calculs. Une salle particulière est affectée aux calcula-
teurs chargés de la réduction des anciennes observations
de la lune, car ce travail, fait sur des fonds spéciaux, est
tout à fait en dehors des occupations ordinaires de l'observatoire. Ajouterai-je que dans cette paisible retraite
de Greenwich, séparée du monde par trois lieues d'inter-
valle, chacun semble avoir oublié les affaires de la terre
pour celles du ciel: là aucune main, aucune tête n'est
oisive ; chacun a son emploi et semble concourir avec
zèle à la perfection de l'ensemble. Qu'il me soit donc

permis de remercier ici publiquement M. Airy, non de l'accueil bienveillant qu'il m'a fait, je devais l'attendre d'un homme de son mérite mais de m'avoir montré un établissement où la science que j'aime avec le plus de passion, est encore dignement représentée et habilement cultivée ; où les fonds qu'un gouvernement libéral lui destine, recoivent une sage direction ; où tout le luxe consiste dans la beauté de instruments, et où enfin il n'y a de place que pour le travail, de distinction que pour le savoir."*

How, then, stands this matter? Let us fix the true meaning of those frequent and loud accusations! Surrounded by their exacting duties, to which their thoughts and unremitting pursuits were adjusted, what course was really required of either of these Astronomers, by the announcement on the part of Mr. Adams of the numerical results of a problem demanding immense labour, and not recognised in our previous science as possible? Mr. Adams had then indeed, clomb an eminence which few reach; but his struggles had not been in the sight of the world; and men knew not the intellectual power he had used in

---

* I confess that my line of defence something resembles the fallacy of the quotation of *general merit* to excuse *particular error*. Not so :—the latter question is inseparable here from the amount of the demands on the time of either Observer.

achieving his triumphs. The graduate of Cambridge had yet his spurs to win: at least I am sure that he of all men will most readily acknowledge, that ere then no authority had grown around his name, which warranted the reception of his assertion on matters like these, as a ground for changing the action of our National Observatory. Is it conceived that the results ought at once to have been published? To this I should suspect that Mr. Adams himself, would, until verification of some sort was obtained, have strongly objected: and no man could then have dreamt of imagining that a prize for which no one had seriously started before, was now the object of a neck and neck race!* Are we told, again, that Mr. Challis or Mr. Airy ought to have placed the accuracy of the solution beyond doubt, by repeating and testing the investigation? Something like this, indeed, ARAGO threw at Sir JOHN HERSCHEL in his angry commentary on our countryman's statement, that the result, as declared by any single calculator, required verification:

* One of the most singular things connected with this history seems to me, that although there is a journal published at Cambridge—rigidly devoted to the diffusion of analytic science, and conducted with great ability, not a trace of what Mr. Adams had been doing is to be found in it.

but this is tantamount to the demand that these public officers abandon their own planned labours —that they resign their time probably for nearly a year. I confess, I do not see that the case, *as it then existed*, demanded, or would at all have authorized any such relinquishment of distinct duty;—to justify which, some additional evidence —some augmentation of probability — was certainly needed.* It is a most grievous and injurious error, however, to suppose that Mr. Airy received the communication from Mr. Adams coolly, or expressed in it no interest. With that earnest and ready frankness, which no slight personal intercourse entitles me to attribute to this eminent Inquirer, when his aid is requested towards worthy ends, he sent Mr. Adams from the resources at Greenwich, whatever he considered might forward his researches; and

---

* I am speaking, be it recollected, of affairs as they stood *previous to the denouement.*—I was much interested lately, on reading somewhere a criticism on Sir John Moore's unfortunate campaign by the Duke of Wellington, to find the high-minded veteran adding,—" *Recollect this is a criticism after the event.*" I once was under a schoolmaster who, beside what he was feed for, instructed me much about politics and the French Revolution. He was, I remember, particularly severe on the unfortunate Moore; but he did not add, " *Recollect this is a criticism after the event!*"

—evidently with the view of obtaining *verifica-tion*—he asked whether his new and remarkable theory explained one peculiar but most important error of Uranus, to which his own thoughts had been especially directed? No answer, it appears, was returned to this question,—probably Mr. Adams had not then carried out his theory in that direction; but whatever the reason, the attempt on the part of the Astronomer Royal to TEST the investigation in an effective way, in the meantime, failed.*

II. If, however, leisure and official avocations prohibited the personal verification of the *theory* of Neptune, ought not some powerful Telescope, directed to the spot indicated by Mr. Adams, to have been employed to search for the lurking planet, and so, as afterwards was accomplished at Berlin by M. Galle, have removed all theoretical uncertainties, and secured the discovery of

---

* It is far from just to interpret this inquiry of Mr. Airy's as a quibble or trap. The results transmitted by Mr. Adams, bore only on one class of the inequalities that had affected Uranus: and Mr. Airy, feeling that the theory, if complete, must explain that other class likewise, naturally and wisely put the question, with a view to the verification of what had been done.

this new orb to the science of England?* Now
there is here also a consideration not observed
on a superficial glance, which, nevertheless, must
be thoroughly weighed previous to a just verdict
in the case. Supposing the problem resolvable and
resolved, the question of chief moment in regard of
the bearing of that solution on the actual search
for the planet, was this: Within *what degree of
proximity to the true place*, has this prediction
probably come; for the extent of the heavens
to be scrutinized was the measure of the labour
involved in such an Inquiry. Now, on this
point, Mr. Airy held very strong opinions,
not sustained by the result. In his letters
to Dr. Hussey and Eugene Bouvard, he ex-
presses, without hesitation, his disbelief in our
power to approach nearly to the place of a disturbing planet — supposing it to exist—by the
facts yet recorded. To Dr. Hussey, for instance, in 1834, he writes as follows: " The

* One cannot indeed avoid greatly regretting, that by some
act of publication, directly or indirectly, opportunity was
not given to the few British observers possessed of large
equatoreals to direct them even at random to that special
region. I cannot imagine that the planet would long have
escaped Dr. Robinson or Mr. Lassell: its *disc* in their great
Telescopes might have saved the toil of that preliminary
mapping undertaken at Cambridge.

state of things is this—the mean motion and
other elements derived from the observations between 1781 and 1825, give considerable errors in
1750, *and nearly the same errors in* 1834, *when
the planet is at nearly the same part of its orbit.*
If the mean motion had been determined by
1750 and 1834, this would have indicated nothing. This does not look like irregular perturbation. . . . . I am sure the place of the disturbing planet could not be determined until the
nature of the irregularity was well determined
from several successive revolutions." That Mr.
Airy fell into this error, ought of itself to con-
vince us that few men at that time would have
escaped it; and I am inclined to hope that
my readers are now prepared to see not the
error merely, but the cause of it. Let it be
recollected that the prediction of Neptune was
not effected, through any complete weighing of
all its actions on Uranus, or of those effects
which completely define its influence, and, therefore, fix its place within our planetary system.
The result came out with what I would call a most
singular accuracy from consideration of the *periodical inequalities* alone—and of a *few* only in this
class; nay, farther, it came out so very strangely,

that—*in consequence of the fortuitous relative positions of the two bodies at the time*—although
the assumed distance of the new orb from the
Sun was in error by six hundred millions of
miles, and its periodic time by half a century,
the longitude, or predicted place on the ecliptic,
was in error by less than one degree!   Now I ask
without hesitation, What physical Astronomer,
who had not gone practically through the inquiry,
and, like Leverrier, and Adams, gained their peculiar confidence from sources not otherwise discern-
ible—could, in such circumstances, have spoken
hopefully of the endeavour?   But, passing for the
moment considerations regarding the rationality
of this *a priori* scepticism, observe its practical
and necessary effect on the action of the Observa-
tories cognizant of the existence of such an Inquiry: for *this*, I beg especially to state, is the
point bearing on the questions raised in respect
of DUTY.   Now the labour which seemed demanded for a practical verification of the prediction was—it must be expressly observed—by
no means what M. Galle underwent, when he
ascertained the planet's existence in one night.
So soon as evidence authorized the search, it
seemed—from *a priori* considerations, which my

reader will now quite understand—expedient to survey at least 30° of the Heavens in length, or a twelfth part of the entire circuit of the ecliptic; and this very laborious task, immediately that circumstances authorized it, was undertaken at Cambridge. About midsummer in 1846, rumours reached London, that a young Frenchman had engaged in the same inquiry as Mr. Adams, and attained independently, almost the same conclusion. The verification previously sought among other sources was thus fully supplied; for in the concurrence of the two investigations all doubt or probability of error disappeared. On that instant then, our activity was awakened; and the men to whom, by accident, the work was intrusted, were quite equal to the task. The great Northumberland telescope at Cambridge—at Greenwich there is no suitable instrument—was devoted to the laborious work, and employed with that quiet persevering energy, distinguishing Mr. Challis.—But be it recollected, AT CAMBRIDGE THERE WAS NO BREMIKER'S CHART. The sky in that region had to be mapped *as the initial step*; and Mr. Challis resolved to include in that map all stars above those of the 11th magnitude. The work proceeded as it ought; but as before Mr. Challis did

not feel himself harnessed for a neck and neck
race. On August 4th, he had mapped down the
planet (Leverrier's *last* paper had not then been
read to the Institute), and on August 12th, an
observation was made which secured the knowledge, that this star was a planet: but the requisite comparison not having then been made, the
announcement of this discovery was postponed,
and the eclât lost through the far easier observation of M. Galle.—Such is this history. For
one, I can see in it no premonition of the
downfal of British science. I see mishap and
difference of opinion; but these are not identical
with negligence of duty. Few living Englishmen
have contributed to the higher physical astronomy
so plentifully and importantly as Mr. AIRY: and
unless on the principle of my old schoolmaster
and Sir John Moore, I may not deduce in the
present instance, failing or dereliction. Mr.
CHALLIS wrought, as usual with him, thoroughly
and steadily, though not as under high pressure;
and he—earliest of all—determined the existence
of the new Orb.

But let squabbles, personal as well as national,
vanish and pass out of sight! Beside questions

like these the ordinary quarrels and jealousies,
either of parties or individuals, ought to have no
place. We are dealing concerning high ordinances of God, invisible to sense, yet descried by
the pure Reason : and—what should gladden the
old and arouse the young—concerning the appearance of two powerful Minds, received, before their
manhood has reached its prime, among the band
of the Illustrious—promising, by rivalry and con-
joint endeavour, to prepare for future ages yet
farther developments of the harmonies and gran-
deur of this Universe. Let us cheer them in
their triumphant course;—Ours, and also Pos-
terity's, will be the possessions they shall achieve!

THE END.